If q is any function of several variables $x, \ldots, z$, then

$$\delta q = \sqrt{\left(\frac{\partial q}{\partial x}\,\delta x\right)^2 + \cdots + \left(\frac{\partial q}{\partial z}\,\delta z\right)^2}$$

(for independent random erro..,

Statistical Definitions (*Chapter 4*)

If $x_1, \ldots, x_N$ denote N separate measurements of one quantity x, then we define:

$$\bar{x} = \frac{1}{N}\sum_{i=1}^{N} x_i = \text{mean}; \qquad \text{(p. 84)}$$

$$\sigma_x = \sqrt{\frac{1}{N-1}\sum (x_i - \bar{x})^2} = \text{standard deviation}; \qquad \text{(p. 87)}$$

$$\sigma_{\bar{x}} = \frac{\sigma_x}{\sqrt{N}} = \text{standard deviation of mean or SDOM.} \qquad \text{(p. 89)}$$

The Normal Distribution (*Chapter 5*)

For any limiting distribution $f(x)$ for measurement of a continuous variable x:

$f(x)\,dx$ = probability that any one measurement will give an answer between x and $x + dx$; (p. 106)

$\int_a^b f(x)\,dx$ = probability that any one measurement will give an answer between $x = a$ and $x = b$; (p. 106)

$\int_{-\infty}^{\infty} f(x)\,dx = 1$ is the normalization condition. (p. 107)

The normal distribution is

$$f_{X,\sigma}(x) = \frac{1}{\sigma\sqrt{2\pi}}\,e^{-(x-X)^2/2\sigma^2}, \qquad \text{(p. 112)}$$

where

X = center of distribution
= true value of x
= mean after many measurements,

σ = width of distribution
= standard deviation after many measurements.

The probability of a measurement within t standard deviations of X is

$$P(\text{within } t\sigma) = \frac{1}{\sqrt{2\pi}}\int_{-t}^{t} e^{-z^2/2}\,dz = \text{normal error integral}; \qquad \text{(p. 116)}$$

in particular

$$P(\text{within } 1\sigma) = 68\%.$$

A Series of Books in Physics

Eugene D. Commins, Editor

An Introduction to Error Analysis

The Study of Uncertainties in Physical Measurements

John R. Taylor

Professor of Physics
University of Colorado

University Science Books
Oxford University Press

University Science Books
20 Edgehill Road
Mill Valley, CA 94941

Library of Congress Catalog Card Number: 81-51269

ISBN clothbound: 0-935702-07-5
paperbound: 0-935702-10-5

Printed in the United States of America
10 9 8 7 6 5 4

To My Wife

Preface

All measurements, however careful and scientific, are subject to some uncertainties. Error analysis is the study and evaluation of these uncertainties, its two main functions being to allow the scientist to estimate how large his uncertainties are, and to help him to reduce them when necessary. The analysis of uncertainties, or "errors," is a vital part of any scientific experiment, and error analysis is therefore an important part of any college course in experimental science. It can also be one of the most interesting parts of the course. The challenges of estimating uncertainties and of reducing them to a level that allows a proper conclusion to be drawn can turn a dull and routine set of measurements into a truly interesting exercise.

This book is an introduction to error analysis for use with an introductory college course in experimental physics of the sort usually taken by freshmen or sophomores in the sciences or engineering. I certainly do not claim that error analysis is the most (let alone the only) important part of such a course, but I have found that it is often the most abused and neglected part. In many such courses, error analysis is "taught" by handing out a couple of pages of notes containing a few formulas, and the student is then expected to get on with the job solo. The result is that error analysis becomes a meaningless ritual, in which the student adds a few lines of calculation to the end of each laboratory report, not because he or she understands why, but simply because the instructor has said to do so.

I wrote this book with the conviction that any student, even one who has never heard of the subject, should be able to learn what error analysis is, why it is interesting and important, and how to use the basic tools of the subject in laboratory reports. Part I of the book (Chapters 1 to 5) tries to do all this, with many examples of the kind of experiment encountered in teaching laboratories. The student who masters this material should then know and understand almost all the error analysis he or she would be expected to learn in a freshman laboratory course: error propagation, the use of elementary statistics, and their justification in terms of the normal distribution.

Part II contains a selection of more advanced topics: least-squares fitting, the correlation coefficient, the χ^2 test, and others. These would almost certainly not be included officially in a freshman laboratory course, although a few students might become interested in some of them. However, several of these topics would be needed in a second laboratory course, and it is primarily for that reason that I have included them.

I am well aware that there is all too little time to devote to a subject like error analysis in most laboratory courses. At the University of Colorado we give a one-hour lecture in each of the first six weeks of our freshman laboratory course. These lectures, together with a few homework assignments using the problems at the ends of the chapters, have let us cover Chapters 1 through 4 in detail and Chapter 5 briefly. This gives the students a working knowledge of error propagation and the elements of statistics, plus a nodding acquaintance with the underlying theory of the normal distribution.

From several students' comments at Colorado, it was evident that the lectures were an unnecessary luxury for at least some of the students, who could probably have learned the necessary material from assigned reading and problem sets. I certainly believe the book could be studied without any help from lectures.

Part II could be taught in a few lectures at the start of a second-year laboratory course (again supplemented with some assigned problems). But, even more than Part I, it was intended to be read by the student at any time that his or her own needs and interests might dictate. Its seven chapters are almost completely independent of one another, in order to encourage this kind of use.

I have included a selection of problems at the end of each chapter; the reader does need to work several of these to master the techniques. Most calculations of errors are quite straightforward. A student who finds himself or herself doing many complicated calculations (either in the problems of this book or in laboratory reports) is almost certainly doing something in an unnecessarily difficult way. In order to give teachers and readers a good choice, I have included many more problems than the average reader need try. A reader who did one-third of the problems would be doing well.

Inside the front and back covers are summaries of all the principal formulas. I hope the reader will find these a useful reference, both while studying the book and afterward. The summaries are organized by chapters, and will also, I hope, serve as brief reviews to which the reader can turn after studying each chapter.

Within the text, a few statements—equations and rules of procedure—have been highlighted by a shaded background. This highlighting is reserved for statements that are important and are in their final form (that is, will not be modified by later work). You will definitely need to remember these statements; so they have been highlighted to bring them to your attention.

The level of mathematics expected of the reader rises slowly through the book. The first two chapters require only algebra; Chapter 3 requires differentiation (and partial differentiation in Section 3.9, which is optional); Chapter 5 needs a knowledge of integration and the exponential function. In Part II, I assume that the reader is entirely comfortable with all these ideas.

The book contains numerous examples of physics experiments, but an understanding of the underlying theory is not essential. Furthermore, the examples are mostly taken from elementary mechanics and optics, to make it more likely that the student will already have studied the theory. The reader who needs it can find an account of the theory by looking at the index of any introductory physics text.

Error analysis is a subject about which people feel passionately, and no single treatment can hope to please everyone. My own prejudice is that, when a choice has to be made between ease of understanding and strict rigor, a physics text should choose the former. For example, on the controversial question of combining errors in quadrature versus direct addition, I have chosen to treat direct addition first, since the student can easily understand the arguments that lead to it.

In the last few years, a dramatic change has occurred in student laboratories with the advent of the pocket calculator. This has a few unfortunate consequences—most notably, the atrocious habit of quoting ridiculously *insignificant* figures just because the calculator produced them—but it is from almost every point of view a tremendous advantage, especially in error analysis. The pocket calculator allows one to compute, in a few seconds, means and standard deviations that previously would have taken hours. It renders unnecessary many tables, since one can now compute functions like the Gauss function more quickly than one could find them in a book of tables. I have tried to exploit this wonderful tool wherever possible.

It is my pleasure to thank several people for their helpful comments and suggestions. A preliminary edition of the book was used at several colleges, and I am grateful to many students and colleagues for their criticisms. Especially helpful were the comments of John Morrison and David Nesbitt at the University of Colorado, Professors Pratt and Schroeder at Michigan State, Professor Shugart at U. C. Berkeley, and

Professor Semon at Bates College. Diane Casparian, Linda Frueh, and Connie Gurule typed successive drafts beautifully and at great speed. Without my mother-in-law, Frances Kretschmann, the proofreading would never have been done in time. I am grateful to all of these people for their help; but above all I thank my wife, whose painstaking and ruthless editing improved the whole book beyond measure.

J. R. Taylor
November 1, 1981
Boulder, Colorado

Contents

PART I

An Introduction to Error Analysis

Part I

1. Preliminary Description of Error Analysis
2. How to Report and Use Uncertainties
3. Propagation of Uncertainties
4. Statistical Analysis of Random Uncertainties
5. The Normal Distribution

Part I introduces the basic ideas of error analysis as they are needed in a typical first-year, college physics laboratory. The first two chapters describe what error analysis is, why it is important, and how it can be used in a typical laboratory report. Chapter 3 describes error propagation, whereby uncertainties in one's original measurements "propagate" through calculations to cause uncertainties in one's calculated final answers. Chapters 4 and 5 introduce the statistical methods with which the so-called random uncertainties can be evaluated.

CHAPTER 1

Preliminary Description of Error Analysis

Error analysis is the study and evaluation of uncertainty in measurement. Experience has shown that no measurement, however carefully made, can be completely free of uncertainties. Since the whole structure and application of science depends on measurements, it is therefore crucially important to be able to evaluate these uncertainties and to keep them to a minimum.

In this first chapter we describe some simple measurements that illustrate the inevitable occurrence of experimental uncertainties and show the great importance of knowing how large these uncertainties are. We shall then describe how (in some simple cases, at least) the magnitude of the experimental uncertainties can be realistically estimated, often by means of little more than ordinary common sense.

1.1. Errors as Uncertainties

In science the word "error" does not carry the usual connotations of "mistake" or "blunder." "Error" in a scientific measurement means the inevitable uncertainty that attends all measurements. As such, errors are not mistakes; you cannot avoid them by being very careful. The best you can hope to do is to ensure that errors are as small as reasonably possible, and to have some reliable estimate of how large they are. Most textbooks introduce additional definitions of "error," and we shall discuss some of these later. For the moment, however, we shall use "error" exclusively in the sense of "uncertainty," and treat the two words as being interchangeable.

1.2. Inevitability of Uncertainty

To illustrate the inevitable occurrence of uncertainties, we have only to examine carefully any everyday measurement. Consider, for example, a carpenter who must measure the height of a doorway in order to install a door. As a first rough measurement, he might simply look at the doorway and estimate that it is 210 cm high. This crude "measurement" is certainly subject to uncertainty. If pressed, the carpenter might express this uncertainty by admitting that the height could be as little as 205 or as much as 215 cm.

If he wanted a more accurate measurement, he would use a tape measure, and he might find that the height is 211.3 cm. This measurement is certainly more precise than his original estimate, but it is obviously still subject to some uncertainty, since it is *inconceivable* that he could know the height to be exactly 211.3000 rather than 211.3001 cm, for example.

There are many reasons for this remaining uncertainty, several of which we will be discussing in this book. Some of these causes of uncertainty could be removed if he took enough trouble. For example, one source of uncertainty might be that poor lighting is making it difficult to read the tape; this could be corrected by improving the lighting.

On the other hand, some sources of uncertainty are intrinsic to the process of measurement and can never be entirely removed. For example, let us suppose the carpenter's tape is graduated in half-centimeters. The top of the door will probably not coincide precisely with one of the half-centimeter marks, and if it does not, then the carpenter must *estimate* just where the top lies between two marks. Even if the top happens to coincide with one of the marks, the mark itself is perhaps a millimeter wide; so he must estimate just where the top lies within the mark. In either case, the carpenter ultimately must estimate where the top of the door lies relative to the markings on his tape, and this necessity causes some uncertainty in his answer.

By buying a better tape with closer and finer markings, the carpenter can reduce his uncertainty, but he cannot eliminate it entirely. If he becomes obsessively determined to find the height of the door with the greatest precision that is technically possible, he could buy an expensive laser interferometer. But even the precision of an interferometer is limited to distances of the order of the wavelength of light (about 0.5×10^{-6} meters). Although he would now be able to measure the height with fantastic precision, he still would not know the height of the doorway *exactly*.

Furthermore, as our carpenter strives for greater precision, he will encounter an important problem of principle. He will certainly find that the height is different in different places. Even in one place, he will find that the height varies if the temperature and humidity vary, or even if he accidentally rubs off a thin layer of dirt. In other words, he will find that there is no such thing as *the* height of the doorway. This kind of problem is called a *problem of definition* (the height of the door is not a well-defined quantity) and plays an important role in many scientific measurements.

Our carpenter's experiences illustrate what is found to be generally true. No physical quantity (a length, a time, a temperature, etc.) can be measured with complete certainty. With care we may be able to reduce the uncertainties until they are extremely small, but to eliminate them entirely is impossible.

In everyday measurements we do not usually bother to discuss uncertainties. Sometimes the uncertainties simply are not interesting. If we say that the distance between home and school is 3 miles, it does not matter (for most purposes) whether this means "somewhere between 2.5 and 3.5 miles" or "somewhere between 2.99 and 3.01 miles." Often the uncertainties are important, but can be allowed for instinctively and without explicit consideration. When our carpenter comes to fit his door, he must know its height with an uncertainty that is less than 1 mm or so. However, as long as the uncertainty is this small, the door will (for all practical purposes) be a perfect fit, and his concern with error analysis is at an end.

1.3. Importance of Knowing the Uncertainties

Our example of the carpenter measuring a doorway illustrated how there are always uncertainties in measurements. We will now consider an example that illustrates more clearly the crucial importance of knowing how big these uncertainties are.

Suppose we are faced with a problem like the one said to have been solved by Archimedes. We are asked to find our whether a crown is made of 18-karat gold, as claimed, or is a cheaper alloy. Following Archimedes, we decide to test the crown's density, knowing that the densities of 18-karat gold and the suspected alloy are

$$\rho_{\text{gold}} = 15.5 \text{ gm/cm}^3$$

and

$$\rho_{\text{alloy}} = 13.8 \text{ gm/cm}^3.$$

If we can measure the density ρ_{crown} of the crown, then it should be possible (as Archimedes suggested) for us to decide whether the crown is really gold, by comparing ρ_{crown} with the known densities ρ_{gold} and ρ_{alloy}.

Suppose we summon two experts in the measurement of density. The first expert, A, might make a quick measurement of ρ_{crown} and report that his best estimate for ρ_{crown} is 15, and that ρ_{crown} almost certainly lies somewhere between 13.5 and 16.5 gm/cm^3. Expert *B* might take a little longer, and then report a best estimate of 13.9 and a probable range from 13.7 to 14.1 gm/cm^3. The findings of our two experts can be summarized as shown in Table 1.1.

Table 1.1. Density of crown (in gm/cm^3).

Measurement reported	Expert *A*	Expert *B*
Best estimate for ρ_{crown}	15	13.9
Probable range for ρ_{crown}	13.5 to 16.5	13.7 to 14.1

The first point to notice about these results is that although *B*'s measurement is much more precise, *A*'s measurement is probably correct also. Each expert states a range within which he is confident ρ_{crown} lies, and these ranges overlap; so it is perfectly possible (and in fact probable) that both statements are correct.

The next point to notice is that the uncertainty in *A*'s measurement is so large that his results are of no use. The densities of 18-karat gold and of the alloy both lie in his range, 13.5 to 16.5 gm/cm^3; so it is impossible to draw any conclusion from *A*'s measurements. On the other hand, *B*'s measurements indicate clearly that the crown is not genuine; the density of the suspected alloy, 13.8, lies comfortably inside *B*'s estimated range of 13.7 to 14.1, but that of 18-karat gold, 15.5, is well outside it. Evidently, if the measurements are to allow a conclusion, the experimental uncertainties must not be too large. However, it is *not* necessary that the uncertainties be extremely small. In this respect our example is typical of many scientific measurements, where uncertainties have to be reasonably small (perhaps a few percent of the measured value), but where extreme precision is often quite unnecessary.

Since our decision hinges on *B*'s claim that ρ_{crown} lies between 13.7 and 14.1 gm/cm^3, it is important that *B* give us sufficient reason to believe his claim. In other words, the experimenter must justify his stated range of values. This point is often overlooked by the beginning student, who simply asserts that his uncertainty was 1 mm, or 2 sec, or whatever, omitting any justifications. Without a brief explanation of how the uncertainty was estimated, the assertion is almost useless.

The most important point about our two experts' measurements is this: like most scientific measurements, they would both have been useless, if they had not included reliable statements of their uncertainties. In fact, if we knew only the information on the top line of Table 1.1, not only would we have been unable to draw any valid conclusion, but we could actually have been misled, since expert A's result (15) would seem to suggest that the crown is genuine.

1.4. More Examples

The examples in the last two sections were chosen because they give a good introduction to some of the principal features of error analysis. They were not chosen for their great importance, and the reader can be excused for thinking them a little contrived. It is, however, easy to think of examples that are of the greatest importance in almost any branch of applied or basic science.

In the applied sciences, the engineer designing a nuclear power plant must know the characteristics of the materials and fuels that he plans to use. The manufacturer of a pocket calculator must know the properties of its various electronic components. In each case, somebody must measure the required parameters; having measured them, he must establish the reliability of his measurements, and this requires error analysis. Engineers concerned with the safety of airplanes, trains, or cars must understand the uncertainties in drivers' reaction times, in braking distances, and in a host of other variables; and failure to carry out error analysis can lead to accidents like that shown on the cover of this book. Even in a less scientific field, like the manufacture of clothing, error analysis, in the form of *quality control*, plays a vital part.

In the basic sciences, error analysis has an even more fundamental role. When any new theory is proposed, it must be tested against older theories by means of one or more experiments for which the new and old theories predict different outcomes. In principle, one simply performs the experiment and lets the outcome decide between the rival theories. In practice, the situation is complicated by the inevitable experimental uncertainties. These must all be carefully analyzed and their effects reduced until the experiment singles out just one acceptable theory. That is, the experimental results, with their uncertainties, must be *consistent* with the predictions of one theory, and *inconsistent* with those of all known, reasonable alternatives. Obviously the success of such a procedure depends critically on the scientist's understanding of error analysis and on his ability to convince others of his understanding.

A famous example of this kind of test of a scientific theory is the measurement of the bending of light when it passes near the Sun. When Einstein published his general theory of relativity in 1916, he pointed out that the theory predicted that light from a star would be bent through an angle $\alpha = 1.8''$ as it passes near the Sun. The simplest classical theory would predict no bending ($\alpha = 0$), and a more careful classical analysis predicts (as Einstein himself had pointed out in 1911) bending through an angle $\alpha = 0.9''$. In principle, all that was necessary was to observe a star when it was aligned with the edge of the Sun, and to measure the angle of bending α. If the result were $\alpha = 1.8''$, then general relativity would be vindicated (at least for this phenomenon); if α were found to be 0 or $0.9''$, then general relativity would be wrong, and one of the older theories right.

In practice, measurement of the bending of light by the Sun was extremely hard, and was only possible during a solar eclipse. Nonetheless, in 1919 it was successfully measured by Dyson, Eddington, and Davidson, who reported their best estimate as $\alpha = 2''$, with 95 percent confidence that it lay somewhere between $1.7''$ and $2.3''$.[1] Obviously this result was consistent with general relativity and inconsistent with either of the older predictions. Therefore it gave strong support to Einstein's theory of general relativity.

At the time this result was controversial. Many people suggested that the uncertainties had been badly underestimated, and hence that the experiment was inconclusive. Subsequent experiments have tended to confirm Einstein's prediction and to vindicate the conclusion of Dyson, Eddington, and Davidson. The important point here is that the whole question hinged on the experimenters' ability to estimate reliably all their uncertainties and to convince everyone else that they had done so.

The student in the introductory physics laboratory will not usually be able to conduct definitive tests of new theories. On the other hand, many experiments in the introductory laboratory are designed as tests of existing physical theories. For example, Newton's theory of gravity predicts that bodies fall with constant acceleration g (under the appropriate conditions), and the student can conduct experiments to test whether this prediction is correct. At first sight, this kind of experiment may seem artificial and pointless, since the theories have obviously been tested many times with much more precision than is possible in any teaching laboratory. Nonetheless, if the student understands the crucial role of error analysis and accepts the challenge to make the most precise test possible with the available equipment, such experiments can be interesting and instructive exercises.

[1] This simplified account is based on the original paper of Dyson, Eddington, and Davidson (*Philosophical Transactions of the Royal Society*, **220A**, 1920, 291). I have converted the *probable error* originally quoted into the 95 percent confidence limits. The precise significance of such confidence limits will be established in Chapter 5.

1.5. Estimating Uncertainties When Reading Scales

So far we have considered several examples that illustrate why every measurement suffers from uncertainties and why it is important to know their magnitude. On the other hand, we have not yet discussed how one can actually evaluate the magnitude of an uncertainty. In fact, such evaluation can be fairly complicated, and is the main topic of the rest of this book. Fortunately, there are some simple measurements for which it is easy to make a reasonable estimate of the uncertainty, often using no more than common sense. Here and in Section 1.6 we give two examples of such simple measurements. An understanding of these examples will allow the student to begin using error analysis in his experiments, and will form the basis for our later developments.

Our first example is a measurement using a marked scale, such as the ruler in Figure 1.1 or the voltmeter in Figure 1.2. To measure the length of the pencil in Figure 1.1, we must first place the end of the pencil opposite the zero of the ruler, and then decide where the tip comes to on the ruler's scale. To measure the voltage in Figure 1.2, we have to decide where the needle points on the voltmeter's scale. If we assume that the ruler and

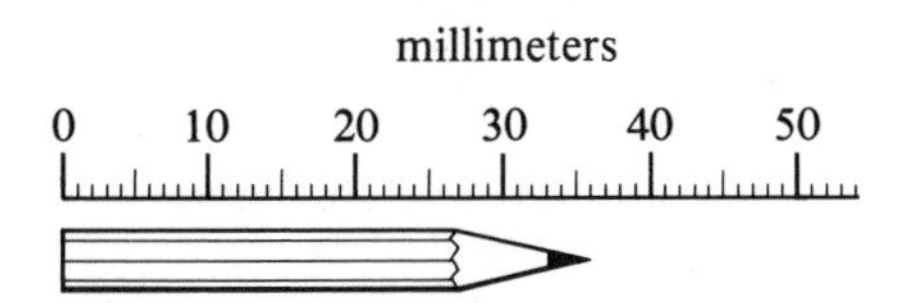

Figure 1.1. Measuring a length with a ruler.

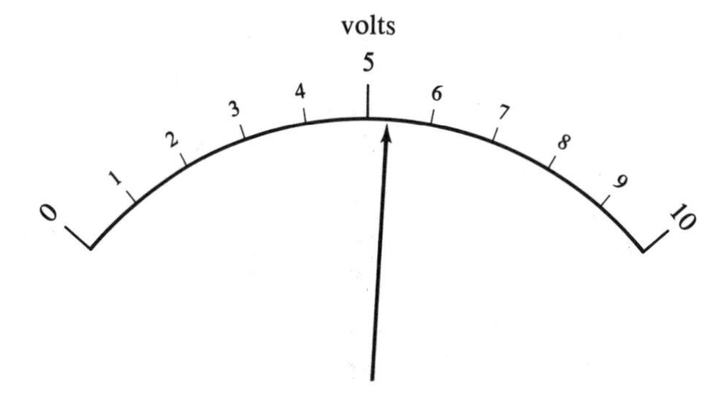

Figure 1.2. A reading on a voltmeter.

voltmeter are reliable, then in each case the main problem is to decide where a certain point lies in relation to the scale markings. (Of course, if there is any possibility that the ruler and voltmeter are *not* reliable, then we shall have to take this into account as well.)

The markings of the ruler in Figure 1.1 are fairly close together (one millimeter apart). An experimenter might reasonably decide that the length shown is undoubtedly closer to 36 mm than it is to 35 or 37 mm, but that no more precise reading is possible. He would therefore state his conclusion as

$$\begin{aligned} &\text{best estimate of length} = 36 \text{ mm}, \\ &\text{probable range,} \qquad 35.5 \text{ to } 36.5 \text{ mm}, \end{aligned} \tag{1.1}$$

and would say that he has measured the length to the nearest millimeter.

This type of conclusion—that the quantity lies closer to one given mark than to either of its neighboring marks—is quite common. For this reason, many scientists introduce the convention that the statement $l =$ 36 mm without any qualification is presumed to mean that l is closer to 36 than to 35 or 37; that is,

$$l = 36 \text{ mm}$$

means

$$35.5 \text{ mm} \leqslant l \leqslant 36.5 \text{ mm}.$$

In the same way, an answer like $x = 1.27$ without any stated uncertainty would be presumed to mean that x lies between 1.265 and 1.275. In this book we will not be using this convention; instead, we will always indicate our uncertainties explicitly. Nevertheless, it is important for the student to understand the convention, and to know that it applies to any number stated without an uncertainty. It is especially important to know about this convention in this age of pocket calculators, which often display many digits. If a student blindly copies a number like 123.456 from his calculator without any qualification, then his reader is entitled to assume the number is definitely correct to six significant figures, which is very unlikely to be so.

The markings on the voltmeter shown in Figure 1.2 are more widely spaced than those on the ruler. Here most observers would agree that one can do more than simply identify the mark to which the pointer is closest. Because the spacing is larger, one can realistically estimate where the pointer lies in the space between two marks. Thus a reasonable conclusion for the voltage shown might be

$$\begin{aligned} &\text{best estimate of voltage} = 5.3 \text{ volts,} \\ &\text{probable range,} \quad 5.2 \text{ to } 5.4 \text{ volts.} \end{aligned} \tag{1.2}$$

The process of estimating positions between the scale markings is called *interpolation.* It is an important technique that can be improved with practice.

Different observers might not agree with the precise estimates given in Equations (1.1) and (1.2). In particular, one might well decide that one could interpolate for the length in Figure 1.1, and measure it with a smaller uncertainty than that given in Equation (1.1). Nevertheless, few people would deny that Equations (1.1) and (1.2) are *reasonable* estimates of the quantities concerned and of their probable uncertainties. Thus we see that approximate estimation of uncertainties is fairly easy when the only problem is to locate a point on a marked scale.

1.6. Estimating Uncertainties in Repeatable Measurements

Many measurements involve uncertainties that are much harder to estimate than those connected with locating points on a scale. For example, when we measure a time interval using a stopwatch, the main source of uncertainty is not the difficulty of reading the dial, but our own unknown reaction time in starting and stopping the watch. These kinds of uncertainty can sometimes be reliably estimated, if we can repeat the measurement several times. Suppose, for example, we time the period of a long pendulum once and get an answer of 2.3 seconds. From one measurement we can't say much about the experimental uncertainty. But if we repeat the measurement and get 2.4 sec, then we can immediately say that the uncertainty is probably of the order of 0.1 sec. If a sequence of four timings gives the results (in sec),

$$2.3,\ 2.4,\ 2.5,\ 2.4, \tag{1.3}$$

then we can begin to make some fairly realistic estimates.

First, it is natural to assume that the best estimate of the period is the *average value*, 2.4 sec.[2]

[2] We will prove in Chapter 5 that the best estimate based on several measurements of a quantity is almost always the average of the measurements.

Second, it seems reasonably safe to assume that the correct period lies somewhere between the lowest value, 2.3, and the highest, 2.5. Thus we might reasonably conclude that

$$\begin{aligned} &\text{best estimate} = \text{average} = 2.4 \text{ sec}, \\ &\text{probable range,} \qquad 2.3 \text{ to } 2.5 \text{ sec}. \end{aligned} \tag{1.4}$$

Whenever we can repeat the same measurement several times, the spread in our measured values gives a valuable indication of the uncertainty in our measurements. In Chapters 4 and 5, we will discuss statistical methods for treating such repeated measurements. Under the right conditions, these statistical methods give a more accurate estimate of uncertainty than we have found in Equation (1.4) using just common sense. A proper statistical treatment also has the advantage of giving an objective value for the uncertainty, independent of the observer's individual judgment.[3] Nevertheless, the estimate in (1.4) represents a simple and realistic conclusion to draw from the four measurements in (1.3).

Repeated measurements such as those in (1.3) cannot always be relied on to reveal the uncertainties. First, we must be sure that the quantity measured is really the *same* quantity each time. Suppose, for example, we measure the breaking strength of two supposedly identical wires by breaking them (something we can't do more than once with each wire). If we get two different answers, this difference *may* indicate that our measurements were uncertain *or* that the two wires were not really identical. By itself, the difference between the two answers sheds no light on the reliability of our measurements.

Even when we can be sure that we are measuring the same quantity each time, repeated measurements will not always reveal uncertainties. For example, suppose that the clock used for the timings in (1.3) was running consistently 5 percent fast. Then all timings made with it will be 5 percent too long, and no amount of repeating (with the same clock) will reveal this deficiency. Errors of this sort, which affect all measurements in the same way, are called *systematic* errors, and can be hard to detect, as we will discuss in Chapter 4. In our example, the remedy would be to check the clock against a more reliable one. More generally, it should be clear that if one has reason to doubt the reliability of any measuring device (a clock, a measuring tape, a voltmeter), one should try to check it against a device that is known to be more reliable.

[3] Also, a proper statistical treatment usually gives a *smaller* uncertainty than the full range from the lowest to the highest observed value. Thus, looking at the four timings in (1.3), we have judged that the period is "probably" somewhere between 2.3 and 2.5 sec. The statistical methods of Chapters 4 and 5 let us state with 70 percent confidence that it lies in the smaller range from 2.36 to 2.44 sec.

The examples discussed in this and the previous section show that experimental uncertainties can sometimes be easily estimated. On the other hand, there are many measurements for which the uncertainties are *not* so easily evaluated. Also, we will ultimately want more precise values for the uncertainties than the simple estimates just discussed can give us. These topics will occupy us from Chapter 3 onward. In Chapter 2 we will assume temporarily that we know how to estimate the uncertainties in all quantities of interest, so that we can discuss how the uncertainties are best reported, and how they are used in drawing an experimental conclusion.

CHAPTER 2

How to Report and Use Uncertainties

We now have some idea of the importance of experimental uncertainties and of how they arise. We have also seen how they can be estimated in some simple situations. In this chapter we present some basic concepts and rules of error analysis, and give examples of their use in some typical experiments in a physics laboratory. Our main aim is to familiarize you with the basic vocabulary of error analysis and with the way it is used in the introductory laboratory. Then, starting in Chapter 3, we will be ready to study how uncertainties are actually evaluated.

In Sections 2.1 to 2.3 we define several basic concepts in error analysis, and discuss some general rules for stating uncertainties. In Sections 2.4 to 2.6 we discuss how these ideas could be used in some typical experiments in an introductory physics laboratory. Finally, in Sections 2.7 to 2.9, we introduce one more basic definition, that of the fractional uncertainty, and discuss its significance.

2.1. Best Estimate ± Uncertainty

We have seen that the correct way to state the result of any measurement is for the measurer to give his best estimate of the quantity concerned and the range within which he is confident it lies. For example, the result of the timings discussed in Section 1.6 was reported as

$$\begin{aligned} &\text{best estimate of time} = 2.4 \text{ sec,} \\ &\text{probable range,} \qquad 2.3 \text{ to } 2.5 \text{ sec.} \end{aligned} \tag{2.1}$$

Here the best estimate, 2.4 sec, lies at the midpoint of the estimated range of probable values, 2.3 to 2.5 sec, as it has in all our examples. This relationship is obviously very natural, and is what pertains in almost all measurements. It allows one to express the results of the measurement in a very compact form. For example, the measurement of the time recorded in (2.1) is usually stated as follows:

$$\text{measured value of time} = 2.4 \pm 0.1 \text{ sec.} \tag{2.2}$$

This single equation is exactly equivalent to the two statements in (2.1).

In general, the result of any measurement of a quantity x is stated as

$$(\text{measured value of } x) = x_{\text{best}} \pm \delta x. \tag{2.3}$$

This statement means, first, that the experimenter's best estimate for the quantity concerned is the number x_{best}, and, second, that he is reasonably confident the quantity lies somewhere between $x_{\text{best}} - \delta x$ and $x_{\text{best}} + \delta x$. The number δx is called the *uncertainty*, or *error*, in the measurement of x. It is convenient always to define the uncertainty δx to be positive, so that $x_{\text{best}} + \delta x$ is always the *highest* probable value of the measured quantity and $x_{\text{best}} - \delta x$ the *lowest*.

We have intentionally left the meaning of the range, $x_{\text{best}} - \delta x$ to $x_{\text{best}} + \delta x$, somewhat vague. We can sometimes make it more precise. In a simple measurement, like that of the height of a doorway, we can easily state a range $x_{\text{best}} - \delta x$ to $x_{\text{best}} + \delta x$ within which we are *absolutely* certain the measured quantity lies. Unfortunately, in most scientific measurements it is very hard to make such a statement. In particular, if we wish to be *completely* certain that the measured quantity lies between $x_{\text{best}} - \delta x$ and $x_{\text{best}} + \delta x$, we usually have to choose a value for δx that is too large to be practical. To avoid this, we can sometimes choose a δx for which we are, for example, 70 percent confident that the actual quantity lies between $x_{\text{best}} - \delta x$ and $x_{\text{best}} + \delta x$. However, we obviously cannot do this without a detailed knowledge of the statistical laws that govern the process of measurement. We will return to this point in Chapter 4. For now, let us content ourselves with defining the uncertainty δx so that we are "reasonably certain" the measured quantity lies somewhere between $x_{\text{best}} - \delta x$ and $x_{\text{best}} + \delta x$.

2.2. *Significant Figures*

Several basic rules for stating uncertainties are worth emphasizing. First, since the quantity δx is an estimate of an uncertainty, it should obviously not be stated with too much precision. If we measure the acceleration of gravity g, it would be absurd to state a result like

$$(\text{measured } g) = 9.82 \pm .02385 \text{ m/sec}^2. \tag{2.4}$$

It is inconceivable that the uncertainty in the measurement can be known to four significant figures. In high-precision work, uncertainties are sometimes stated with two significant figures, but for the introductory laboratory we can state the following rule.[1]

Rule for Stating Uncertainties

In an introductory laboratory, experimental uncertainties should usually be rounded to one significant figure. (2.5)

Thus, if some calculation yields the uncertainty $\delta g = .02385 \text{ m/sec}^2$, this answer should be rounded to $\delta g = .02 \text{ m/sec}^2$, and the conclusion (2.4) should be rewritten as

$$(\text{measured } g) = 9.82 \pm .02 \text{ m/sec}^2. \tag{2.6}$$

An important practical consequence of this rule is that many error calculations can be carried out in one's head, without the aid of a calculator, or even of pencil and paper.

There is only one significant exception to the rule in (2.5). If the leading digit in the uncertainty δx is a 1, then it may be better to keep two significant figures in δx. For example, suppose that some calculation gave the uncertainty $\delta x = .14$. To round this to $\delta x = .1$ would be a 40 percent reduction; so one could argue that it might be less misleading to retain two figures, and quote $\delta x = .14$. The same argument could conceivably be applied if the leading digit is a 2, but certainly not if it is any larger.

Once the uncertainty in a measurement has been estimated, one must also consider which are the significant figures in the measured value. A statement like

$$\text{measured speed} = 6051.78 \pm 30 \text{ m/sec} \tag{2.7}$$

is obviously ridiculous. The uncertainty of 30 means that the digit 5 in the third place of 6051.78 might really be as small as 2 or as large as 8.

[1] For convenience in referring to these rules, they have been included in the numbering sequence for equations; some of them do and some do not contain equations.

Clearly the trailing digits 1, 7, and 8 have no significance at all, and should be rounded off. That is, the correct statement of (2.7) is

$$\text{measured speed} = 6050 \pm 30 \text{ m/sec}. \tag{2.8}$$

Clearly the general rule is as follows.

Rule for Stating Answers

The last significant figure in any stated answer should usually be of the same order of magnitude (in the same decimal position) as the uncertainty. (2.9)

For example, the answer 92.81 with an uncertainty of .3 should be rounded as

$$92.8 \pm .3.$$

If its uncertainty is 3, then the same answer should be rounded as

$$93 \pm 3,$$

and if the uncertainty is 30, then the answer should be

$$90 \pm 30.$$

However, numbers to be used in calculations should generally be kept with *one more significant figure* than is finally justified. This will reduce the inaccuracies introduced by rounding the numbers. At the end of the calculation, the final answer should be rounded to remove this extra (and insignificant) figure.[2]

Note that the uncertainty in any measured quantity has the same dimensions as the measured quantity itself. It is therefore clearer and more economical to write the units (m/sec^2, cm^2, etc.) after both the answer

[2] There is one more small exception to the rule in (2.9). If the leading digit in the uncertainty is small (a 1 or perhaps a 2), then it may be appropriate to retain one extra figure in the final answer. For example, an answer such as

$$\text{measured length} = 27.6 \pm 1 \text{ cm}$$

is quite acceptable, since one could argue that to round it to 28 $\pm$ 1 cm would be to waste information.

and the uncertainty, as in Equations (2.6) and (2.8). By the same token, if a measured number is so large or small that it calls for "scientific notation" (i.e., the use of the form 3×10^3 instead of 3,000), then it is simpler and clearer to put the answer and uncertainty in the same form. For example, the result

$$\text{measured charge} = (1.61 \pm .05) \times 10^{-19} \text{ coulombs}$$

is much easier to read and understand in this form than it would be in the form

$$\text{measured charge} = 1.61 \times 10^{-19} \pm 5 \times 10^{-21} \text{ coulombs.}$$

2.3. *Discrepancy*

Before we address the question of how to use uncertainties in experimental reports, a few important terms need to be introduced and defined. First, if two measurements of the same quantity disagree, then we say that there is a *discrepancy*. Numerically, we define the discrepancy between the two measurements as their difference:

$$\text{discrepancy} = \text{difference between two measured values of the same quantity.} \tag{2.10}$$

It is important to recognize that a discrepancy may or may not be significant. If two students measure the same resistance and get the answers

$$40 \pm 5 \text{ ohms}$$

and

$$42 \pm 8 \text{ ohms,}$$

the discrepancy of 2 ohms is less than their uncertainties; so the two measurements are obviously consistent. Here we would say that the discrepancy is *insignificant*. On the other hand, if the two answers had been

$$35 \pm 2 \text{ ohms}$$

and

$$45 \pm 1 \text{ ohms},$$

then the two measurements would be clearly inconsistent, and the discrepancy of 10 ohms would be *significant*. Here some careful checks would be needed to discover what had gone wrong.

In the teaching laboratory, one frequently measures quantities (like c, the speed of light, or e, the electron's charge) that have been accurately measured many times before, and for which a very accurate *accepted value* is known and published in the handbooks. This accepted value is not, of course, exact; it is the result of measurements and, like all measurements, has some uncertainty. Nonetheless, in many cases the accepted value is much more accurate than the student can possibly achieve himself. For example, the currently accepted value of c, the speed of light, is

$$(\text{accepted } c) = 299{,}792{,}458 \pm 1 \text{ m/sec}. \tag{2.11}$$

As expected, this value *is* uncertain, but the uncertainty is extremely small by the standards of most teaching laboratories.[3]

Although there are many experiments in which one measures a quantity whose accepted value is known, there are very few measurements for which the *true answer* is known.[4] In fact, the true value of a measured quantity can almost *never* be known exactly and is, in fact, hard to define. Nevertheless, it is sometimes useful to discuss the difference between some measured value and the corresponding true value, and some authors call this difference the *true error*.

2.4. Comparison of Measured and Accepted Values

There is very little point in performing an experiment if one does not draw some sort of conclusion. A very few experiments may have mainly qualitative results—the appearance of an interference pattern on a ripple

[3] This is not always so. For example, if one looks up the refractive index of glass, one finds values ranging from 1.5 to 1.9, depending on the composition of the glass. In an experiment to measure the refractive index of a piece of glass whose composition is unknown, the "accepted" value is therefore no more than a very rough guide to the expected answer.

[4] Since the reader may have trouble in thinking of *any* such measurements, here is an example. If one measures the ratio of a circle's circumference to its diameter, the true answer is exactly π. Obviously such experiments are very contrived.

tank or the color of light transmitted by some optical system—but the vast majority of experiments lead to *quantitative* conclusions, that is, to a statement of numerical results. It is therefore important to recognize that the statement of *a single measured number is completely uninteresting.* Statements that the density of some metal was measured as 9.3 ± 0.2 gm/cm^3 or that the momentum of a cart was measured as 0.051 ± 0.004 kg·m/sec are, by themselves, of no interest. An interesting conclusion must *compare two or more numbers*: a measurement with the accepted value; a measurement with a theoretically predicted value; or several measurements, to show that they are related to one another in accordance with some physical law. It is in such comparison of numbers that error analysis is so important. In this and the next two sections, we discuss three typical experiments to illustrate how the estimated uncertainties are used to draw a conclusion.

Perhaps the simplest type of experiment is a measurement of a quantity whose accepted value is known. As we have already discussed, this is a somewhat artificial experiment, peculiar to the teaching laboratory. In it one measures the quantity, estimates one's experimental uncertainty, and finally compares these with the accepted value. Thus an experiment to measure the speed of sound in air (at standard temperature and pressure) might arrive at the conclusion that

$$\text{measured speed} = 329 \pm 5 \text{ m/sec}, \tag{2.12}$$

compared with the

$$\text{accepted speed} = 331 \text{ m/sec}. \tag{2.13}$$

To this numerical conclusion the student would probably add the comment that since the accepted speed lies inside the estimated range of the measured speed, the measurement was satisfactory; and his or her report could be complete.

The meaning of the uncertainty δx is that the correct value of x "probably" lies between $x_{\text{best}} - \delta x$ and $x_{\text{best}} + \delta x$; it is certainly *possible* that the correct value lies slightly outside this range. Therefore a measurement can be regarded as satisfactory even if the accepted value lies slightly outside the estimated range of the measured value. For example, a measured value of 325 ± 5 m/sec can be considered compatible with the accepted value of 331 m/sec. On the other hand, if the accepted value is well outside the measured range (discrepancy much more than twice the uncertainty, say), then there is good reason to think something has gone wrong. Thus the unlucky student who finds

$$\text{measured speed} = 345 \pm 2 \text{ m/sec} \tag{2.14}$$

compared with the

$$\text{accepted speed} = 331 \text{ m/sec} \tag{2.15}$$

is going to have to check his measurements and calculations to find out what has gone wrong.

Unfortunately, the tracing of his mistake can be a tedious business, since there are so many possibilities. He may have made a mistake in the measurements or calculations that led to the answer 345 m/sec. He may have estimated his uncertainty incorrectly. (The answer 345 ± 10 m/sec would have been acceptable.) He might be comparing his measurement with the wrong accepted value. For example, the accepted value 331 m/sec is the speed of sound at standard temperature and pressure. Since standard temperature is 0°C, there is a good chance that the measured speed in (2.14) was *not* measured at standard temperature. In fact, if the measurement was made at 20°C (i.e., normal room temperature), then the correct accepted value for the speed of sound is 343 m/sec, and the measurement would be entirely acceptable.

Finally, and perhaps most likely, a discrepancy like that between (2.14) and (2.15) may indicate some undetected source of systematic error (like a clock that is running consistently slow, as discussed in Chapter 1). The detection of such systematic errors (ones that consistently push the result in one direction) will require careful checking of the calibration of all instruments and detailed review of all procedures.

2.5. Comparison of Two Measured Numbers

In many experiments one measures two numbers which theory predicts should be equal. For example, the law of conservation of momentum states that the total momentum of an isolated system is constant. To test it, we might perform a series of experiments with two carts that collide as they move along a frictionless track. We could measure the total momentum of the two carts before they collide (p) and again after the collision (p'), and then check whether $p = p'$ within experimental uncertainties. For a single pair of measurements, our results could be

$$\text{initial momentum } p = 1.49 \pm .04 \text{ kg·m/sec}$$

and

$$\text{final momentum } p' = 1.56 \pm .06 \text{ kg·m/sec.}$$

Here the range in which p probably lies (1.45 to 1.53) *overlaps* the range in which p' probably lies (1.50 to 1.62). Therefore this measurement is consistent with conservation of momentum. If, on the other hand, the two probable ranges were not even close to overlapping, then the measurement would be inconsistent with conservation of momentum; and we would have to check for mistakes in our measurements or calculations, for possible systematic errors, and for the possibility that some external forces (such as gravity or friction) are causing the momentum of the system to change.

Suppose we repeat similar pairs of measurements several times. What is the best way to display our results? First, it is almost always best to record a sequence of similar measurements in a table, not as several distinct statements. Second, our uncertainty often differs only very little from one measurement to the next. For example, we may convince ourselves that the uncertainty in all measurements of the initial p is $\delta p \approx$ .04 kg·m/sec, and that the uncertainty in the final p' is $\delta p' \approx$ .06 kg·m/sec. If this is so, a good way to display our measurements would be as shown in Table 2.1.

Table 2.1. Measured momenta (all in kg·m/sec).

Initial p (all $\pm$.04)	Final p' (all $\pm$.06)
1.49	1.56
2.10	2.12
1.16	1.05
etc.	etc.

For each pair of measurements, the probable range of values of p overlaps (or nearly overlaps) the range of values of p'. If this continues to be true for all measurements, then our results can be pronounced consistent with conservation of momentum.

With a little thought, we can display our results in a way that makes our conclusion even clearer. For example, conservation of momentum requires that the *difference* $p - p'$ be zero. If we add to our table a column showing $p - p'$, then the entries in this column should all have values consistent with zero. The only difficulty with this is that we must know how to compute the uncertainty in the difference $p - p'$. This can be easily done. Suppose we have made measurements

$$(\text{measured } p) = p_{\text{best}} \pm \delta p$$

and

$$(\text{measured } p') = p'_{\text{best}} \pm \delta p'.$$

The numbers p_{best} and p'_{best} are our best estimates for p and p'. Therefore the best estimate for the difference $(p - p')$ is $(p_{\text{best}} - p'_{\text{best}})$. To find the uncertainty in $(p - p')$, we must decide on the highest and lowest probable values of $(p - p')$. The highest value for $(p - p')$ would result if p had its *largest* probable value, $p_{\text{best}} + \delta p$, at the same time that p' had its *smallest* value $p'_{\text{best}} - \delta p'$. Thus the highest probable value for $p - p'$ is

$$\text{highest probable value} = (p_{\text{best}} - p'_{\text{best}}) + (\delta p + \delta p'). \qquad (2.16)$$

Similarly, the lowest probable value arises when p is smallest $(p_{\text{best}} - \delta p)$, but p' is largest $(p'_{\text{best}} + \delta p')$. This gives

$$\text{lowest probable value} = (p_{\text{best}} - p'_{\text{best}}) - (\delta p + \delta p'). \qquad (2.17)$$

Combining (2.16) and (2.17), we see that the *uncertainty in the difference* $(p - p')$ *is the* ***sum*** $\delta p + \delta p'$ *of the original uncertainties.* For example, if

$$p = 1.49 \pm .04 \text{ kg·m/sec}$$

and

$$p' = 1.56 \pm .06 \text{ kg·m/sec}$$

then

$$p - p' = -.07 \pm .1 \text{ kg·m/sec}.$$

We can now add an extra column for $p - p'$ to Table 2.1, and arrive at Table 2.2.

Table 2.2. Measured momenta (all in kg·m/sec).

Initial p (all $\pm .04$)	Final p' (all $\pm .06$)	Difference $p - p'$ (all $\pm .1$)
1.49	1.56	$-.07$
2.10	2.12	$-.02$
1.16	1.05	.11
etc.	etc.	etc.

Whether our results are consistent with conservation of momentum can now be seen at a glance by checking whether the numbers in the final column are consistent with zero (that is, are less than, or comparable with, the uncertainty .1). Another way to achieve the same effect would be to tabulate the *ratios* p'/p, which should all be consistent with the value $p'/p = 1$. (Here we would need to calculate the uncertainty in p'/p, a problem we will take up in Chapter 3.)

Our discussion of the uncertainty in $p - p'$ obviously applies to the difference of any two measured numbers. We have therefore established the following general rule.

Uncertainty in a Difference

If the quantities x and y are measured with uncertainties δx and δy, and if the measured values of x and y are used to calculate the difference $q = x - y$, then the *uncertainty in* q is the *sum of the uncertainties in* x *and* y:

$$\delta q \approx \delta x + \delta y. \tag{2.18}$$

We have used the approximate equality sign ($\approx$) to emphasize two points. First, we do not yet have a precise definition for the uncertainties involved; so it would be absurd to claim that δq is *exactly* equal to $\delta x + \delta y$. Second, we shall see in Section 3.4 that the uncertainty δq is often somewhat less than is given by (2.18); a better estimate is the so-called "quadratic sum" of δx and δy, defined in (3.13). Thus the sign "$\approx$" in (2.18) is used as a reminder that we will be replacing (2.18) with a better estimate later.

The result (2.18) is the first in a series of rules for the *propagation of errors.* When we calculate a quantity q in terms of measured quantities x and y, we need to know how the uncertainties in x and y "propagate" to cause uncertainty in q. We will give a complete discussion of error propagation in Chapter 3.

2.6. Checking Proportionality with a Graph

Many physical laws imply that one quantity should be proportional to another. Hooke's law states that the extension of a spring is proportional to the force stretching it; Newton's law says that the acceleration of a body is proportional to the total applied force; and these are just two of countless examples. Many experiments in a teaching laboratory are designed to check this kind of proportionality.

When one quantity, y, is proportional to some other, x, a graph of y against x is a straight line through the origin. Thus one can test whether

y is proportional to x by plotting the measured values of y against those of x and seeing whether the resulting points do lie on a straight line through the origin. Because a straight line is so easily recognizable, this is a simple and effective way to check for proportionality.

To illustrate this use of graphs, let us imagine an experiment to test Hooke's law. This law, usually written as $F = kx$, asserts that the extension x of a spring is proportional to the force F which is stretching it, $x = F/k$, where k is the "force constant" of the spring. A simple way to test this is to hang the spring vertically and to suspend various masses m from it. Here the force F is the weight mg of the load; so the extension should be

$$x = \frac{mg}{k} = \left(\frac{g}{k}\right)m. \tag{2.19}$$

The extension x should be proportional to the load m, and a graph of x against m should be a straight line through the origin.

If we measure x for a variety of different loads m and plot our measured values of x and m, it is most unlikely that the resulting points will lie *exactly* on a straight line. Suppose, for example, we measure the extension x for eight different loads m and get the results shown in Table 2.3. These values are plotted in Figure 2.1(a), where we also show a possible straight line that passes through the origin and is reasonably close to all eight points. As we should have expected, the eight points do not lie exactly on any line. The question is whether this is just because of experimental uncertainties (as we would hope), or because we have made mistakes, or even because the extension x is *not* proportional to m. To decide this, we must consider our uncertainties.

As usual, the measured quantities, the extensions x and masses m, are subject to some uncertainty. For simplicity let us suppose first that the masses used are known very accurately, so that the uncertainty in m is negligibly small. Suppose, on the other hand, that all measurements of x have an uncertainty of about 0.3 cm (as indicated in Table 2.3). For a load of 200 gm, for example, the extension would therefore probably be in the range 1.1 ± 0.3 cm. Our first experimental point on the graph then lies on

Table 2.3. Load and extension.

Load m (gm) (δm negligible)	200	300	400	500	600	700	800	900
Extension x (cm) (all ± 0.3)	1.1	1.5	1.9	2.8	3.4	3.5	4.6	5.4

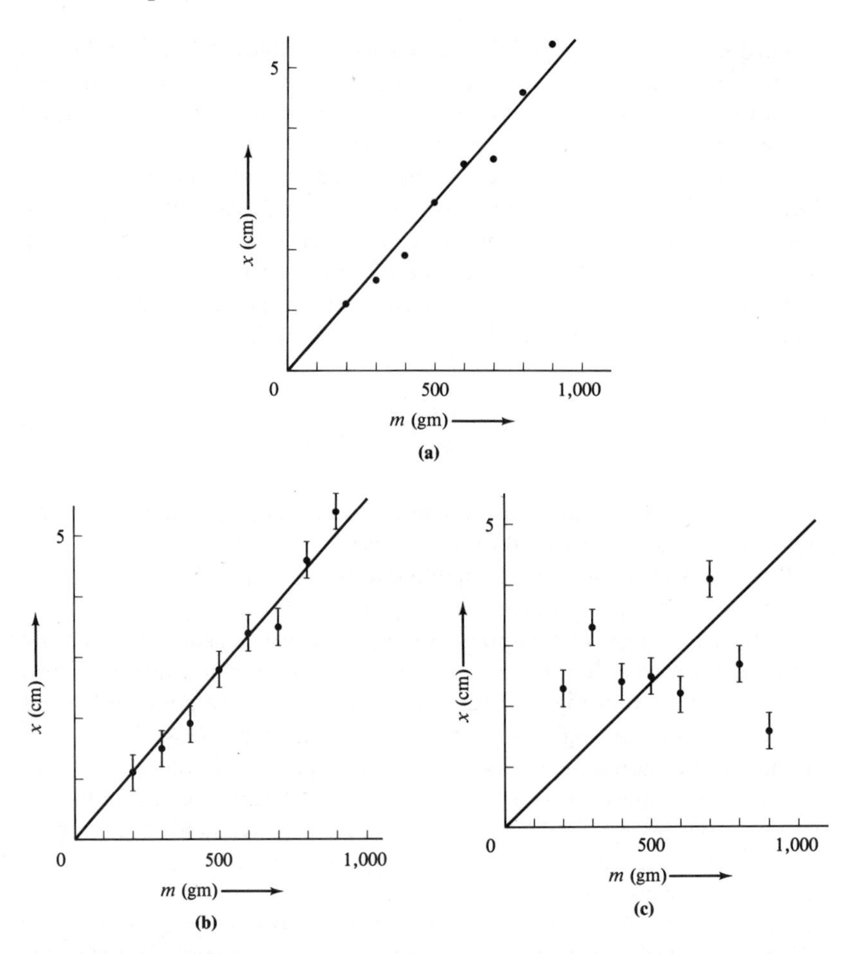

Figure 2.1. Three plots of extension x of a spring against the load m. (a) The data of Table 2.3 without error bars. (b) The same data with error bars to show the uncertainties in x. (The uncertainties in m are assumed to be negligible.) These data are consistent with the expected proportionality of x and m. (c) A different set of data, which are inconsistent with x's being proportional to m.

the vertical line $m = 200$ gm, somewhere between $x = 0.8$ and $x = 1.4$ cm. This is indicated in Figure 2.1(b), where we have drawn a vertical *error bar* through each point to indicate the range in which it probably lies. Obviously we should expect to be able to find a straight line that goes through the origin and *passes through or close to all the error bars*. In Figure 2.1(b) there is such a line; so we would conclude that the data on which Figure 2.1(b) is based are consistent with x's being proportional to m.

We saw in Equation (2.19) that the slope of the graph of x against m is g/k. By measuring the slope of the line in Figure 2.1(b), we can therefore find the constant k of the spring. By drawing the steepest and least steep lines that seem to fit the data reasonably well, we could also find the uncertainty in this value for k. (See Problem 2.8.)

If the best straight line misses a high proportion of the error bars, or if it misses any by a large distance (compared to the length of the error bars), then our results would be *inconsistent* with x's being proportional to m. This situation is illustrated in Figure 2.1(c). With the results shown there, we would have to recheck our measurements and calculations (including the calculation of the uncertainties) and to consider whether there is some reason that x is *not* proportional to m.

So far we have supposed that there is negligible uncertainty in the mass (which is plotted along the horizontal axis), and that the only uncertainties are in x, as shown by vertical error bars. If both x and m are subject to appreciable uncertainties, there are various ways to display them. The simplest is to draw both vertical and horizontal error bars through each point, the length of each arm being equal to the corresponding uncertainty, as shown in Figure 2.2. Each cross in this picture corresponds to one measurement of x and m, with x probably lying in the interval defined by the vertical bar of the cross, and m probably in that defined by the horizontal bar.

A slightly more complicated situation arises when one physical quantity is expected to be proportional to some power of another quantity. Consider the distance x traveled by a body in a time t when falling freely. This distance is $x = \frac{1}{2}gt^2$ and is proportional to the square of t. If we plot x against t, the experimental points should lie on a parabola. However, it is

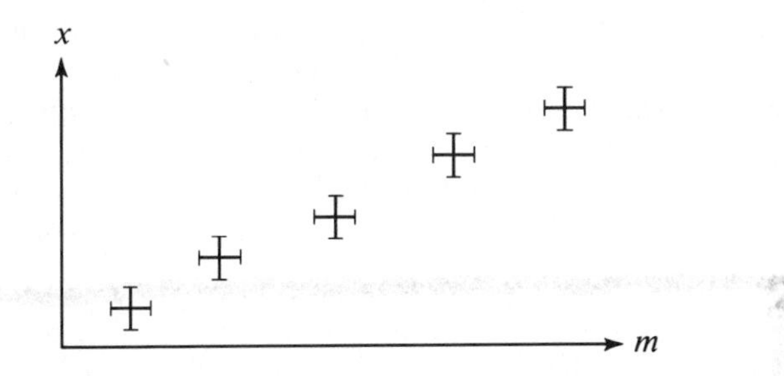

Figure 2.2 Measurements with uncertainties in x and m can be shown by crosses made of one error bar for x and one for m.

hard to check visually whether or not a set of points lies on a parabola (or any other curve except a straight line). A much better procedure is to observe that if $x \propto t^2$, then a graph of x against t^2 should be a straight line through the origin; and such a graph will allow one to check whether or not the data are consistent with a straight line, just as in the preceding example. Similarly, if x is proportional to an exponential function e^{At} (where A is some constant), then a graph of $\ln x$ against t should be a straight line; and such a graph allows one to check easily whether $x \propto e^{At}$. (We discuss this point further in Chapter 8.)

There are other, nongraphical methods for checking the proportionality of two quantities. For example, if $x \propto m$, then the ratio x/m should be constant. One could simply add, to the table of values of x and m, an extra row or column showing the ratios x/m, and thus easily check whether these ratios are constant within their estimated uncertainties. Again, with a programmable calculator, one can write a program that will automatically check how well a set of measurements fits a straight line. However, even when some other method is used to check that $x \propto m$, it is an excellent idea to make the graphical check as well. Graphs like those of Figures 2.1(b) and (c) show clearly how well the measurements verify the predictions; and drawing such graphs helps one understand the experiment and the physical laws involved.

2.7. *Fractional Uncertainties*

The uncertainty δx in a measurement,

$$(\text{measured } x) = x_{\text{best}} \pm \delta x,$$

indicates the reliability or precision of the measurement. However, the uncertainty δx by itself does not tell the whole story. An uncertainty of one inch in a distance of one mile would indicate an unusually precise measurement, whereas an uncertainty of one inch in a distance of three inches would indicate a rather crude estimate. Obviously the quality of a measurement is indicated not just by the uncertainty δx, but also by the *ratio* of δx to x_{best}; and this leads us to consider the *fractional uncertainty*,

$$\text{fractional uncertainty} = \frac{\delta x}{|x_{\text{best}}|}. \tag{2.20}$$

(The fractional uncertainty is also called the *relative uncertainty* or the *precision.*) In this definition, the symbol $|x_{\text{best}}|$ denotes the absolute value[5] of x_{best}.

To avoid confusion with the fractional uncertainty, the uncertainty δx itself is sometimes called the *absolute uncertainty.*

In most serious measurements, the uncertainty δx is much smaller than the measured value x_{best}. Since the fractional uncertainty $\delta x/|x_{\text{best}}|$ is therefore usually a small number, it is often convenient to multiply it by 100 and quote it as the *percentage uncertainty.* For example, the measurement

$$\text{length } l = 50 \pm 1 \text{ cm} \tag{2.21}$$

has a fractional uncertainty

$$\frac{\delta l}{|l_{\text{best}}|} = \frac{1}{50} = .02$$

and a percentage uncertainty of 2 percent. Thus the result (2.21) could be given as

$$\text{length } l = 50 \text{ cm} \pm 2\%.$$

Be sure to notice that, whereas the absolute uncertainty δl has the same units as l, the fractional uncertainty $\delta l/|l_{\text{best}}|$ is a *dimensionless* quantity, without units. Keeping this difference in mind can help you avoid the common mistake of confusing absolute uncertainty with fractional uncertainty.

The fractional uncertainty is an approximate indication of the quality of a measurement, whatever the size of the quantity measured. Fractional uncertainties of 10 percent or so are usually characteristic of fairly rough measurements. (A rough measurement of 10 inches might have an uncertainty of 1 inch; a rough measurement of 10 miles might have an uncertainty of 1 mile.) Fractional uncertainties of 1 or 2 percent are characteristic of fairly accurate measurements, and are about the best one can hope for in many experiments in the introductory physics laboratory. Fractional uncertainties much less than 1 percent are usually hard to achieve, and are rare in the introductory laboratory.

[5] The absolute value $|x|$ of a number x is equal to x when x is positive, but is obtained by omitting the minus sign if x is negative. We use the absolute value in (2.20) to guarantee that the fractional uncertainty, like the uncertainty δx itself, is always positive, whether x_{best} is positive or negative. In practice, one usually arranges matters so that measured numbers are positive, and the absolute-value signs in (2.20) can then be omitted.

These divisions are, of course, extremely rough. A few very simple measurements can have fractional uncertainties of 0.1 percent or less with very little trouble. With a good tape measure a distance of 10 feet can easily be measured with an uncertainty of $\frac{1}{10}$ inch, or about 0.1 percent; with a good timer, a period of an hour can easily be measured to better than a second, or 0.03 percent. On the other hand, for many quantities that are very hard to measure a 10 percent uncertainty would be regarded as an experimental triumph. Large percentage uncertainties therefore do *not* mean that a measurement is scientifically useless. In fact, many important measurements in the history of physics had experimental uncertainties of 10 percent or more. Certainly there is plenty to be learned in the introductory physics laboratory from equipment whose minimum uncertainty is a few percent.

2.8. *Significant Figures and Fractional Uncertainties*

The concept of fractional uncertainty is closely related to the familiar notion of significant figures. In fact, the number of significant figures in a quantity is an approximate indicator of the fractional uncertainty in that quantity. For example, consider the two numbers

$$510 \quad \text{and} \quad .51,$$

both of which have been certified accurate to two significant figures. Since 510 (with two significant figures) means

$$510 \pm 5 \quad \text{or} \quad 510 \pm 1\%,$$

and .51 means

$$.51 \pm .005 \quad \text{or} \quad .51 \pm 1\%,$$

we see that both numbers are 1 percent uncertain. In other words, the statement that the numbers 510 and .51 have two significant figures is equivalent to saying that they are 1 percent uncertain. In the same way, 510 with three significant figures would have a fractional uncertainty of 0.1 percent, and so on.

Unfortunately, this useful connection is only approximate. The number 110, given to two significant figures, means

$$110 \pm 5 \qquad 110 \pm 5\%,$$

whereas 910 (again with two significant figures) means

$$910 \pm 5 \quad \text{or} \quad 910 \pm .5\%.$$

We see that the fractional uncertainty associated with 2 significant figures varies from .5% to 5% depending on the first digit of the number concerned. For a summary, see Table 2.4.

Table 2.4. Approximate correspondence between significant figures and fractional uncertainties.

Number of significant figures	Corresponding fractional uncertainty	
	is between	or, very roughly, is
1	5% & 50%	10%
2	.5% & 5%	1%
3	.05% & .5%	.1%

2.9. *Multiplying Two Measured Numbers*

What is perhaps the greatest importance of fractional errors emerges when we start multiplying measured numbers by each other. For example, in order to find the momentum of a body, we might measure its mass m and its velocity v, and then multiply them to give the momentum $p = mv$. Both m and v will be subject to uncertainties, which we will have to estimate. The problem, then, is to find the uncertainty in p that results from the known uncertainties in m and v.

First, for convenience, let us rewrite the standard form

$$(\text{measured value of } x) = x_{\text{best}} \pm \delta x$$

in terms of the fractional uncertainty, as

$$(\text{measured value of } x) = x_{\text{best}}\left(1 \pm \frac{\delta x}{|x_{\text{best}}|}\right). \tag{2.22}$$

For example, if the fractional uncertainty is 3 percent, then from (2.22) we see that

$$(\text{measured value of } x) = x_{\text{best}}\left(1 \pm \frac{3}{100}\right);$$

that is, a 3 percent uncertainty means that x probably lies somewhere between x_{best} times 0.97 and x_{best} times 1.03,

$$(0.97) \times x_{\text{best}} \leqslant x \leqslant (1.03) \times x_{\text{best}}.$$

We will find this a useful way to think about a measured number that we are going to have to multiply.

Let us now return to our problem of calculating $p = mv$, when m and v have been measured as

$$(\text{measured } m) = m_{\text{best}}\left(1 \pm \frac{\delta m}{|m_{\text{best}}|}\right) \tag{2.23}$$

and

$$(\text{measured } v) = v_{\text{best}}\left(1 \pm \frac{\delta v}{|v_{\text{best}}|}\right). \tag{2.24}$$

Since m_{best} and v_{best} are our best estimates for m and v, our best estimate for $p = mv$ is

$$p_{\text{best}} = (\text{best estimate for } p) = m_{\text{best}}v_{\text{best}}.$$

The largest probable values of m and v are given by (2.23) and (2.24) with the plus signs. Thus the largest probable value for $p = mv$ is

$$(\text{largest value for } p) = m_{\text{best}}v_{\text{best}}\left(1 + \frac{\delta m}{|m_{\text{best}}|}\right)\left(1 + \frac{\delta v}{|v_{\text{best}}|}\right). \tag{2.25}$$

The smallest probable value for p is given by a similar expression with two minus signs. Now, the product of the parentheses in (2.25) can be multiplied out as

$$\left(1 + \frac{\delta m}{|m_{\text{best}}|}\right)\left(1 + \frac{\delta v}{|v_{\text{best}}|}\right) = 1 + \frac{\delta m}{|m_{\text{best}}|} + \frac{\delta v}{|v_{\text{best}}|} + \frac{\delta m}{|m_{\text{best}}|}\frac{\delta v}{|v_{\text{best}}|}. \tag{2.26}$$

Since the two fractional uncertainties $\delta m/|m_{\text{best}}|$ and $\delta v/|v_{\text{best}}|$ are small numbers (a few percent, perhaps), their product is extremely small. Therefore the last term in (2.26) can be neglected. Returning to (2.25) we find

$$(\text{largest value of } p) = m_{\text{best}} v_{\text{best}} \left(1 + \frac{\delta m}{|m_{\text{best}}|} + \frac{\delta v}{|v_{\text{best}}|}\right).$$

The smallest probable value is given by a similar expression with two minus signs. Our measurements of m and v therefore lead to a value of $p = mv$ given by

$$(\text{value of } p) = m_{\text{best}} v_{\text{best}} \left(1 \pm \left[\frac{\delta m}{|m_{\text{best}}|} + \frac{\delta v}{|v_{\text{best}}|}\right]\right).$$

Comparing this with the general form

$$(\text{value of } p) = p_{\text{best}} \left(1 \pm \frac{\delta p}{|p_{\text{best}}|}\right),$$

we see that the best estimate for p is $p_{\text{best}} = m_{\text{best}} v_{\text{best}}$ (as we already knew), and that the *fractional uncertainty in p is the sum of the fractional uncertainties in m and v,*

$$\frac{\delta p}{|p_{\text{best}}|} = \frac{\delta m}{|m_{\text{best}}|} + \frac{\delta v}{|v_{\text{best}}|}.$$

If, for example, we had the following measurements for m and v,

$$m = .53 \pm .01 \; kg$$

and

$$v = 9.1 \pm 0.3 \; m/sec,$$

then the best estimate for $p = mv$ is

$$p_{\text{best}} = m_{\text{best}} v_{\text{best}} = (.53) \times (9.1) = 4.82 \text{ kg·m/sec}.$$

To compute the uncertainty in p, we would first compute the fractional errors

$$\frac{\delta m}{m_{\text{best}}} = \frac{.01}{.53} = .02 = 2\%$$

and

$$\frac{\delta v}{v_{\text{best}}} = \frac{0.3}{9.1} = .03 = 3\%.$$

The fractional uncertainty in p is then the sum:

$$\frac{\delta p}{p_{\text{best}}} = 2\% + 3\% = 5\%.$$

If we want to know the absolute uncertainty in p, we must multiply by p_{best}:

$$\delta p = \frac{\delta p}{p_{\text{best}}} \times p_{\text{best}} = .05 \times 4.82 = 0.241.$$

We then round δp and p_{best} to give us our final answer

$$(\text{value of } p) = 4.8 \pm .2 \text{ kg}\cdot\text{m/sec}.$$

The preceding considerations apply to any product of two measured quantities. We have therefore discovered our second general rule for the propagation of errors. If we measure two quantities and form their product, then the uncertainties in the original two quantities "propagate" to cause an uncertainty in their product. This uncertainty is given by the following rule.

Uncertainty in a Product

If x and y have been measured with small fractional uncertainties $\delta x/|x_{\text{best}}|$ and $\delta y/|y_{\text{best}}|$, and if the measured values of x and y are used to calculate the product $q = xy$, then the *fractional uncertainty in q is the sum of the fractional uncertainties in x and y,*

$$\frac{\delta q}{|q_{\text{best}}|} \approx \frac{\delta x}{|x_{\text{best}}|} + \frac{\delta y}{|y_{\text{best}}|}. \tag{2.27}$$

We have used the approximate equality sign ($\approx$) in (2.27), because, just as with the rule for uncertainty in a difference, we will be replacing (2.27) with a more precise rule later on. Two other features of this rule also need emphasizing. First, the derivation of (2.27) required that the fractional uncertainties in x and y both be small enough that we could neglect their product. This is almost always true in practice; so we will always assume it. Nevertheless, it should be remembered that if the fractional uncertainties are not much smaller than 1, the rule in (2.27) does not apply. Second, even when x and y have different dimensions, Equation (2.27) balances dimensionally, since all fractional uncertainties are dimensionless.

In physics we are continually multiplying numbers together; so the rule (2.27) for finding the uncertainty in a product will obviously be an important tool in error analysis. For the moment our main purpose is to emphasize that the uncertainty in any product $q = xy$ is expressed most simply in terms of fractional uncertainties, as in (2.27).

Problems

Note: An asterisk by a problem indicates that the problem is discussed, or its answer given, in the Answers section at the back of the book.

2.1 (Section 2.1). In Chapter 1 a carpenter reported his measurement of the height of a doorway by stating that his best estimate was 210 cm and that he was confident the height was somewhere between 205 and 215 cm. Rewrite this result in the standard form $x_{\text{best}} \pm \delta x$. Do the same for the measurements reported in Equations (1.1), (1.2), and (1.4).

***2.2** (Section 2.2). Rewrite the following answers in their clearest forms, with a suitable number of significant figures:

(a) measured height $= 5.03 \pm .04329$ meters;
(b) measured time $= 19.5432 \pm 1$ sec;
(c) measured charge $= -3.21 \times 10^{-19} \pm 2.67 \times 10^{-20}$ coulombs;
(d) measured wavelength $= 0.000{,}000{,}563 \pm 0.000{,}000{,}07$ meters;
(e) measured momentum $= 3.267 \times 10^{3} \pm 42$ gm·cm/sec.

***2.3** (Section 2.3).

(a) A student measures the density of a liquid five times and gets the results (all in gm/cm^3), 1.8, 2.0, 2.0, 1.9, 1.8. What would you suggest as the best estimate and uncertainty based on his measurements?
(b) He is told that the accepted value is 1.85 gm/cm^3. What is the discrepancy (between his best estimate and the accepted value)? Do you think it significant?

2.4 (Section 2.5). The time for ten revolutions of a turntable is measured by noting the starting and stopping times with the second hand of a wrist watch and subtracting. If the starting and stopping times are uncertain by ± 1 sec each, what is the uncertainty in the time for ten revolutions?

***2.5** (Section 2.5). In an experiment to check conservation of angular momentum, a student obtains the results shown in Table 2.5 for the initial and final angular momenta (L *and* L') of a rotating system. Add an extra column to Table 2.5 to show the difference $L - L'$ and its uncertainty. Are the student's results consistent with conservation of angular momentum?

Table 2.5. Angular momenta (in kg·m²/sec).

Initial L	Final L'
$3.0 \pm .3$	$2.7 \pm .6$
$7.4 \pm .5$	8.0 ± 1
14.3 ± 1	16.5 ± 1
25 ± 2	24 ± 2
32 ± 2	31 ± 2
37 ± 2	41 ± 2

2.6 (Section 2.5). An experimenter measures the masses M and m of a car and trailer. He gives his results in the standard form $M_{\text{best}} \pm \delta M$ and $m_{\text{hest}} \pm \delta m$. What would be his best estimate for the total mass $M + m$? By considering what the largest and smallest probable values of the total mass are, show that his uncertainty in the total mass is just the sum of δM and δm. State your arguments clearly; don't just write down the answer.

2.7 (Section 2.6). Using the data of Problem 2.5, make a plot of final angular momentum L' against initial angular momentum L for the experiment described there. Include vertical and horizontal error bars. (As with all graphs, label your axes clearly, including units. Use proper graph paper. Choose scales so that your graph fills a reasonable proportion of the paper and, in this case, be sure to include the origin.) On what curve would you expect all points to lie? Do the points lie on the expected curve (within experimental uncertainties)?

***2.8** (Section 2.7). If a stone is thrown upward with velocity v, then it should rise to a height h satisfying the equation $v^2 = 2gh$. In particular, v^2 should be proportional to h. To test this, a student measures v^2 and h for seven different throws, with the results shown in Table 2.6.

(a) Make a plot of v^2 against h, including vertical and horizontal error bars. (As usual, label your axes, use proper squared paper, and choose your scale sensibly.) Is your plot consistent with the prediction that $v^2 \propto h$?

Table 2.6. Heights and speeds.

h in meters (all $\pm .05$)	v^2 in m^2/sec^2
.4	7 ± 3
.8	17 ± 3
1.4	25 ± 3
2.0	38 ± 4
2.6	45 ± 5
3.4	62 ± 5
3.8	72 ± 6

(b) The slope of your graph should be $2g$. To find the slope, draw what seems the best straight line through the origin and all the points, and then measure its slope. To find the uncertainty in the slope, draw the steepest and least steep lines which seem to fit the data reasonably. The slopes of these lines give the largest and smallest probable values of the slope. Are your results consistent with the accepted value $2g = 19.6$ m/sec^2?

***2.9** (Section 2.6).
(a) In an experiment with a simple pendulum, a student decides to check whether the period T is independent of the amplitude A (defined as the largest angle that the pendulum makes with the vertical during its oscillations). He obtains the results shown in Table 2.7. Draw a graph of T against A. (Consider carefully your choice of scales. If you feel any doubt about this, draw two graphs, one including the origin $A = 0$, $T = 0$, and one in which only values of T between 1.9 and 2.2 sec are shown.) Should the student conclude that the period is independent of amplitude?
(b) Discuss how the conclusions to part (a) would be changed if all the measured values of T had been uncertain by ± 0.3 sec.

Table 2.7. Amplitude and period of a pendulum.

Amplitude A (deg)	Period T (sec)
5 ± 2	$1.932 \pm .005$
17 ± 2	$1.94 \pm .01$
25 ± 2	$1.96 \pm .01$
40 ± 4	$2.01 \pm .01$
53 ± 4	$2.04 \pm .01$
67 ± 6	$2.12 \pm .02$

2.10 (Section 2.7). Compute the percentage uncertainties for the five measurements reported in Problem 2.2. (Don't forget to round to a reasonable number of significant figures.)

2.11 (Section 2.7). A meter stick can be read to the nearest mm; a traveling microscope can be read to the nearest .1 mm. Suppose you want to measure a length of 2 cm with a precision of 1 percent. Can you do so with the meter stick? Is it possible to do so with the microscope?

***2.12** (Section 2.7). In order to calculate the acceleration of a cart, a student measures its initial and final velocities, v_i and v_f, and computes the difference $(v_f - v_i)$. His data in two separate trials (all in cm/sec) are shown in Table 2.8. All four measurements have 1 percent uncertainty.

Table 2.8. Initial and final speeds.

	v_i	v_f
First run	14.0	18.0
Second run	19.0	19.6

(a) Calculate the absolute uncertainties in all four measurements; find the change $(v_f - v_i)$ and its uncertainty in each run.
(b) Compute the percentage uncertainty for each of the two values of $(v_f - v_i)$. (Your answers here, particularly for the second run, illustrate the disastrous results of measuring a small number by taking the difference of two much larger numbers.)

2.13 (Section 2.8).
(a) A student's calculator shows an answer 123.123. If the student decides that this number actually has only three significant figures, what are its absolute and fractional uncertainties?
(b) Do the same for the number .123,123.
(c) Do the same for the number 321.321.
(d) Do the fractional uncertainties lie in the range expected for three significant figures?

***2.14** (Section 2.9).
(a) A student measures two quantities a and b with the results $a = 11.5 \pm .2$ cm and $b = 25.4 \pm .2$ cm. He now calculates the product $q = ab$. Find his answer, giving both its percentage uncertainty and its absolute uncertainty.
(b) Repeat part (a) for the measurements $a = 10 \pm 1$ cm and $b = 27.2 \pm .1$ sec.
(c) Repeat part (a) with $a = 3.0$ ft $\pm 8\%$ and $b = 4.0$ lb $\pm 2\%$.

***2.15** (Section 2.9).

(a) A student measures two numbers x and y as

$$x = 10 \pm 1, \qquad y = 20 \pm 1.$$

What is his best estimate for their product $q = xy$? Using the largest probable values for x and y (11 and 21) calculate the highest probable value of q. Similarly, find the smallest probable value of q, and hence the range in which q probably lies. Compare your result with that given by the rule in (2.27).

(b) Do the same for the measurements

$$x = 10 \pm 8, \qquad y = 20 \pm 15.$$

Remember that the rule in (2.27) was derived by assuming that the fractional uncertainties are much less than 1.

2.16 (Section 2.9). A well-known rule states that when two numbers are multiplied together, the answer will be reliable if rounded to the number of significant figures in the less precise of the original two numbers.

(a) Using our rule in (2.27) and the fact that significant figures correspond roughly to fractional uncertainty, prove that this "well-known rule" is *approximately* valid. (To be definite, treat the case that the less-precise number has two significant figures.)

(b) Show by an example that the answer can actually be somewhat less precise than the "well-known rule" suggests. (This is especially true if one multiplies together several numbers.)

CHAPTER 3

Propagation of Uncertainties

Most physical quantities usually cannot be measured in a single direct measurement, but are instead found in two distinct steps. First, one measures one or more quantities $x, y, \ldots$, that *can* be measured directly and from which the quantity of interest can be calculated. Second, using the measured values of $x, y, \ldots$, one calculates the quantity of interest itself. For example, to find the area of a rectangle, one actually measures its length l and height h, and then calculates its area A as $A = lh$. Similarly, the most obvious way to find the velocity v of some body is to measure the distance traveled, d, and the time taken, t, and then to calculate v as $v = d/t$. The reader who has spent any time in an introductory laboratory will have no trouble in thinking up more examples. In fact, a little thought will show that almost all interesting measurements involve these two distinct steps, of direct measurement followed by calculation.

When a measurement involves these two steps, the estimation of uncertainties also involves two steps. One must first estimate the uncertainties in the quantities that are measured directly, and then find out how these uncertainties "propagate" through the calculations to produce an uncertainty in the final answer.[1] This "propagation of errors" is the main subject of this chapter.

In fact, we have already discussed some examples of propagation of errors in Chapter 2. In Section 2.5 we discussed what happens when two numbers x and y are measured and the results used to calculate the difference $q = x - y$. We found that the uncertainty in q is just the *sum* $\delta q \approx \delta x + \delta y$ of the uncertainties in x and y. In Section 2.9 we discussed the product $q = xy$, and in Problem 2.6 we discussed the sum $q = x + y$. We will review these cases in Section 3.2. In the rest of this chapter we will

[1] In Chapter 4 we will discuss another way in which the final uncertainty can sometimes be estimated. If all measurements can be repeated several times, and if one is sure that all uncertainties are random in character, then the uncertainty in the quantity of interest can be estimated by examining the spread in answers. Even when this method is possible, it is usually best used as a check on the two-step procedure discussed in this chapter.

discuss more general cases of propagation of uncertainties and work out several examples.

In Section 3.1, before we take up the subject of error propagation, we will discuss briefly the estimation of uncertainties in quantities that are measured directly. We will review the methods discussed in Chapter 1, and then discuss some further examples of error estimation in direct measurements.

Starting in Section 3.2 we shall take up the propagation of errors. We will find that almost all problems in error propagation can be solved using just three simple rules. We will also state a single, more complicated rule that covers all cases, and from which the three simpler rules can all be derived.

This is a rather long chapter. However, the reader can omit the last two sections without any loss of continuity.

3.1. Uncertainties in Direct Measurements

Almost all direct measurements involve reading a scale (on a ruler, a clock, or a voltmeter, for example) or a digital display (on a digital clock or voltmeter, for example). Some of the problems in scale reading were discussed in Section 1.5. Sometimes the main sources of uncertainty are the reading of the scale and the need to interpolate between the scale markings. In such situations a reasonable estimate of the uncertainty is easily made. For example, if one has to measure some clearly defined length l with a ruler graduated in millimeters, one might reasonably decide that the length could be read to the nearest milimeter but no better. Here the uncertainty δl would be $\delta l = 0.5$ mm. If the scale markings are further apart (as with tenths of an inch), one might reasonably decide that one could read to one-fifth of a division, for example. In any case, the uncertainties associated with the reading of a scale can obviously be estimated quite easily and realistically.

Unfortunately, there are frequently other sources of uncertainty that are much more important than any difficulties in scale reading. In measuring the distance between two points, the main problem may be to decide where those two points really are. For example, in an optics experiment one may wish to measure the distance q from the center of a lens to some focused image, as in Figure 3.1. In practice, the lens is usually several millimeters thick, so that locating its center will be hard; and if the lens comes in a bulky mounting, as it often does, then locating the center will be even harder. Furthermore, the image may appear to be

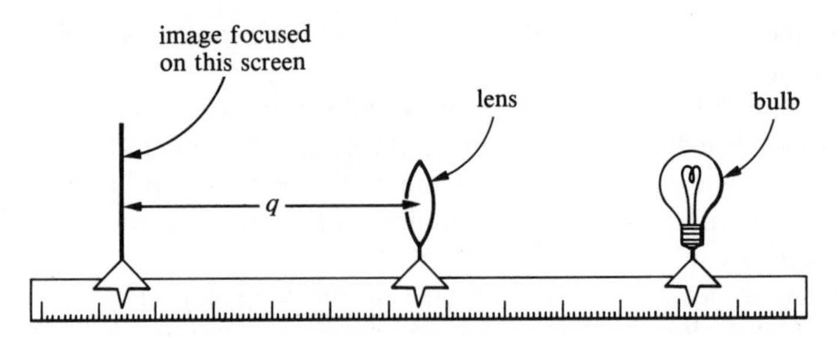

Figure 3.1. An image of the light bulb on the right is focused by the lens onto the screen at the left.

well-focused throughout a range of many millimeters. Even though the apparatus is mounted on an optical bench that is clearly graduated in millimeters, the uncertainty in the distance from lens to image could therefore easily be a centimeter or so. Since this uncertainty arises because the two points concerned are not clearly defined, this kind of problem is called a *problem of definition.*

This example illustrates a serious danger in error estimation. If one looks only at the scales and forgets about other sources of uncertainty, one can underestimate the total uncertainty very badly. In fact, the beginning student's most common mistake is to overlook some sources of uncertainty and hence to *underestimate* uncertainties, often by a factor of 10 or more. However, it is also important not to *overestimate* errors. An experimenter who decides to play safe and to quote generous uncertainties on all measurements may avoid embarrassing inconsistencies, but his or her measurements may not be of much use to anyone. Clearly the ideal is to find all possible causes of uncertainty and to estimate their effects accurately, which is usually not quite so hard as one might suppose.

Reading a digital display is easier than reading a conventional scale. Unless such a device is defective, it will display only significant figures. A digital timer which shows seconds with two decimal places and a time t of 8.03 seconds must (at the worst) mean that

$$t = 8.03 \pm .01 \text{ sec.}$$

Depending on how the counter works, the uncertainty may be half this big. That is, 8.03 may mean $8.03 \pm .005$.

The digital display, even more than the conventional scale, can give a misleading impression of accuracy. For example, one might use a digital timer to time the fall of a weight in an Atwood machine or a similar device. If the timer displays three decimal places and shows a time $t =$ 8.036 sec, then apparently the time of fall is

$$t = 8.036 \pm .001 \text{ sec.} \tag{3.1}$$

However, the careful student who repeats the experiment under conditions as nearly identical as possible might find the second measurement to be

$$t = 8.113 \text{ sec.}$$

One likely explanation of this discrepancy is that uncertainties in the starting procedure are varying the initial conditions and hence the measured time of fall. In any case, it is clear that the accuracy claimed in (3.1) is ridiculously too good. Judging from the two measurements made so far, a more realistic answer would be

$$t = 8.07 \pm .05 \text{ sec.}$$

This example brings us to another point mentioned in Chapter 1. Whenever a measurement can be repeated, it should usually be made several times. The resulting spread of values is often a good indication of the uncertainties, and the average of the values is almost certainly more trustworthy than any one separate measurement. In Chapters 4 and 5 we will discuss the statistical treatment of multiple measurements. Here we wish only to emphasize that if a measurement is repeatable, it should be repeated, both to obtain a more reliable answer (by averaging) and, even more important, to get some estimate of the uncertainties. Unfortunately, as was also mentioned in Chapter 1, repeating a measurement will not always reveal uncertainties. If the measurement is subject to some systematic error, which pushes all results in the same direction (like a clock that runs slow), then the spread in results will not reflect this systematic error. To eliminate such systematic errors will require careful checks of calibration and procedures.

Finally, there is another, quite different kind of measurement, whose uncertainty can be easily estimated. Some experiments involve counting events that occur at random, but with a definite average rate. For example, in a sample of radioactive material, each individual nucleus decays at a random time, but there is a definite average rate at which we would expect to see decays occur in the whole sample. We can try to measure this average rate by observing how many decays occur within some definite interval, such as one minute. (This can be done, for example, by using a Geiger counter to count the charged particles ejected by each nucleus when it decays.) Suppose we find that ν decays have occurred after we have counted for one minute. Because the decays occur at random, we cannot be sure that ν really is the correct average number of decays to be expected in one minute. The question is, of course, how unreliable ν is as a measure of the expected average number of events.

We will discuss the theory of such counting problems in Chapter 11, but the answer is remarkably simple and can be stated now. If we count the number of events in some time T and get the answer ν, then, as a measure of the expected average number of events in the time T, our answer ν has an uncertainty of $\sqrt{\nu}$. That is, our conclusion (based on this one observation) should be

$$\text{(average number of events in time } T) = \nu \pm \sqrt{\nu}. \tag{3.2}$$

For example, if we count 15 decays from a sample of radioactive uranium in one minute, we would conclude that, on average, our sample undergoes $15 \pm \sqrt{15}$, or 15 ± 4, decays per minute.

3.2. *Sums and Differences; Products and Quotients*

In the remainder of this chapter we will suppose that we have measured one or more quantities $x, y, \ldots$, with corresponding uncertainties $\delta x, \delta y, \ldots$, and that we now wish to use the measured values of $x, y, \ldots$, to calculate the quantity of real interest, q. The calculation of q is usually straightforward; the problem we need to discuss is how the uncertainties $\delta x, \delta y, \ldots$, propagate through the calculation and lead to an uncertainty δq in the final value of q.

Sums and Differences

In Chapter 2 we discussed what happens when one measures two quantities x and y, and calculates their sum, $x + y$, or their difference, $x - y$. To estimate the uncertainty in either the sum or difference, we had only to decide on their highest and lowest probable values. The highest and lowest probable values of x are $x_{\text{best}} \pm \delta x$, and those of y are $y_{\text{best}} \pm \delta y$. Hence the highest probable value of $x + y$ is

$$x_{\text{best}} + y_{\text{best}} + (\delta x + \delta y),$$

and the lowest probable value is

$$x_{\text{best}} + y_{\text{best}} - (\delta x + \delta y).$$

Thus the best estimate for $q = x + y$ is

$$q_{\text{best}} = x_{\text{best}} + y_{\text{best}},$$

and its uncertainty is

$$\delta q \approx \delta x + \delta y. \tag{3.3}$$

A similar argument (be sure you can reconstruct it) shows that the uncertainty in the *difference* $x - y$ is given by the same formula (3.3). That is, the uncertainty in either the sum $x + y$ or the difference $x - y$ is the *sum* $\delta x + \delta y$ of the uncertainties in x and y.

If we have several numbers $x, \ldots, w$ to be added or subtracted, then repeated application of (3.3) gives the following rule.

Uncertainty in Sums and Differences

If several quantities $x, \ldots, w$ are measured with uncertainties $\delta x, \ldots, \delta w$, and the measured values used to compute

$$q = x + \cdots + z - (u + \cdots + w),$$

then the uncertainty in the computed value of q is the sum,

$$\delta q \approx \delta x + \cdots + \delta z + \delta u + \cdots + \delta w, \tag{3.4}$$

of all the original uncertainties.

In other words, when one adds or subtracts any number of quantities, the uncertainties in those quantities always *add*. As before, we use the sign $\approx$ to emphasize that we will soon be improving on this rule.

Example

As a simple example of the rule in (3.4), suppose an experimenter mixes together the liquids in two flasks, having first measured their separate masses when full and empty, as follows:

M_1 = mass of first flask and contents $= 540 \pm 10$ gm;
m_1 = mass of first flask empty $= 72 \pm 1$ gm;
M_2 = mass of second flask and contents $= 940 \pm 20$ gm;
m_2 = mass of second flask empty $= 97 \pm 1$ gm.

He now calculates the total mass of liquid as

$$\begin{aligned} M &= M_1 - m_1 + M_2 - m_2 \\ &= (540 - 72 + 940 - 97)\text{ gm} = 1{,}311\text{ gm}. \end{aligned}$$

According to the rule in (3.4), the uncertainty in this answer is the sum of all four uncertainties,

$$\begin{aligned} \delta M \approx \delta M_1 + \delta m_1 + \delta M_2 + \delta m_2 &= (10 + 1 + 20 + 1)\text{ gm} \\ &= 32\text{ gm}. \end{aligned}$$

Thus his final answer (properly rounded) is

$$\text{total mass of liquid} = 1{,}310 \pm 30\text{ gm}.$$

Notice that the much smaller uncertainties in the masses of the empty flasks made a negligible contribution to the final uncertainty. This is an important effect, which we will discuss later on. With experience, the student can learn to identify in advance those uncertainties that are negligible and that can be ignored from the outset. Often this can greatly simplify the calculation of uncertainties.

Products and Quotients

In Section 2.9 we discussed the uncertainty in the product $q = xy$ of two measured quantities. We saw that, provided the fractional uncertainties concerned are small, the *fractional* uncertainty in $q = xy$ is the sum of the *fractional* uncertainties in x and y. Rather than review the derivation of this result, we discuss here the similar case of the quotient $q = x/y$. As we will see, the uncertainty in a quotient is given by the same rule as for a product; that is, the fractional uncertainty in $q = x/y$ is equal to the sum of the fractional uncertainties in x and y.

Since uncertainties in products and quotients are best expressed in terms of fractional uncertainties, it is convenient to introduce a shorthand notation for the latter. Recall that, if we measure some quantity x as

$$(\text{measured value of } x) = x_{\text{best}} \pm \delta x$$

in the usual way, then the fractional uncertainty in x is defined to be

$$(\text{fractional uncertainty in } x) = \frac{\delta x}{|x_{\text{best}}|}.$$

(The absolute value in the denominator ensures that the fractional uncertainty is always positive, even when x_{best} is negative.) Because the symbol $\delta x/|x_{best}|$ is clumsy to write and read, from now on we will abbreviate it by omitting the subscript "best" and writing

$$\text{(fractional uncertainty in } x) = \frac{\delta x}{|x|}.$$

The result of measuring any quantity x can be expressed in terms of its fractional error $\delta x/|x|$ as

$$\text{(value of } x) = x_{best}(1 \pm \delta x/|x|).$$

Therefore the value of $q = x/y$ can be written as

$$\text{(value of } q) = \frac{x_{best}}{y_{best}} \frac{1 \pm \delta x/|x|}{1 \pm \delta y/|y|}.$$

Our problem is now to find the extreme probable values of the second factor on the right. This factor is largest, for example, if the numerator has its largest value, $1 + \delta x/|x|$, and the denominator has its *smallest* value, $1 - \delta y/|y|$. Thus the largest probable value for $q = x/y$ is

$$\text{(largest value of } q) = \frac{x_{best}}{y_{best}} \frac{1 + \delta x/|x|}{1 - \delta y/|y|}. \tag{3.5}$$

The last factor in expression (3.5) has the form $(1 + a)/(1 - b)$, where the numbers a and b are normally small (i.e., much less than 1). It can be simplified by two approximations. First, since b is small, the binomial theorem[2] implies that

$$\frac{1}{(1 - b)} \approx 1 + b. \tag{3.6}$$

Therefore

$$\begin{aligned} \frac{1 + a}{1 - b} &\approx (1 + a)(1 + b) = 1 + a + b + ab \\ &\approx 1 + a + b \end{aligned}$$

[2] The binomial theorem expresses $1/(1 - b)$ as the infinite series $1 + b + b^2 + \cdots$. If b is much less than 1, then $1/(1 - b) \approx 1 + b$ as in (3.6). The reader who is unfamiliar with the binomial theorem can find more details in Problem 3.7.

where, in the second line, we have neglected the product ab of two small quantities. Returning to (3.5) and using these approximations, we find for the largest probable value of $q = x/y$

$$\text{(largest value of } q) = \frac{x_{\text{best}}}{y_{\text{best}}}\left(1 + \frac{\delta x}{|x|} + \frac{\delta y}{|y|}\right).$$

A similar calculation shows that the smallest probable value is given by a similar expression with two minus signs. Combining these, we find that

$$\text{(value of } q) = \frac{x_{\text{best}}}{y_{\text{best}}}\left(1 \pm \left[\frac{\delta x}{|x|} + \frac{\delta y}{|y|}\right]\right).$$

Comparing this with the standard form,

$$\text{(value of } q) = q_{\text{best}}\left(1 \pm \frac{\delta q}{|q|}\right),$$

we see that the best value for q is $q_{\text{best}} = x_{\text{best}}/y_{\text{best}}$, as we would expect, and that the fractional uncertainty is

$$\frac{\delta q}{|q|} \approx \frac{\delta x}{|x|} + \frac{\delta y}{|y|}. \tag{3.7}$$

We conclude that when we divide or multiply two measured quantities x and y, the fractional uncertainty in the answer is the sum of the fractional uncertainties in x and y, as in (3.7). If we now multiply or divide a series of numbers, then repeated application of this result leads to the following general rule.

Uncertainty in Products and Quotients

If several quantities $x, \ldots, w$ are measured with small uncertainties $\delta x, \ldots, \delta w$, and the measured values are used to compute

$$q = \frac{x \times \cdots \times z}{u \times \cdots \times w},$$

then the fractional uncertainty in the computed value of q is the sum,

$$\frac{\delta q}{|q|} \approx \frac{\delta x}{|x|} + \cdots + \frac{\delta z}{|z|} + \frac{\delta u}{|u|} + \cdots + \frac{\delta w}{|w|}, \tag{3.8}$$

of the fractional uncertainties in $x, \ldots, w$.

Briefly, when one multiplies or divides quantities the *fractional uncertainties add.*

Example

In surveying one can sometimes find a value for an inaccessible length l (such as the height of a tall tree) by measuring three other lengths l_1, l_2, l_3 in terms of which

$$l = \frac{l_1 l_2}{l_3}.$$

Suppose that we perform such an experiment, with the results (in feet)

$$l_1 = 200 \pm 2, \qquad l_2 = 5.5 \pm .1, \qquad l_3 = 10.0 \pm .4.$$

Our best estimate for l is

$$l_{\text{best}} = \frac{200 \times 5.5}{10.0} = 110 \text{ ft.}$$

According to (3.8), the fractional uncertainty in this answer is the sum of the fractional uncertainties in l_1, l_2, l_3, which are 1, 2, and 4 percent. Thus

$$\begin{aligned} \frac{\delta l}{l} \approx \frac{\delta l_1}{l_1} + \frac{\delta l_2}{l_2} + \frac{\delta l_3}{l_3} &= (1 + 2 + 4)\% \\ &= 7\%, \end{aligned}$$

and our final answer is

$$l = 110 \pm 8 \text{ ft.}$$

Measured Quantity Times Exact Number

Two important special cases of the rule in (3.8) deserve to be mentioned separately. First, suppose we measure a quantity x, and then use our result to calculate the product $q = Bx$, where the number B has *no uncertainty*. For example, we might measure the diameter of a circle and then calculate its circumference, $c = \pi \times d$; or we might measure the thickness T of 100 identical sheets of paper and then calculate the thickness of a single sheet as $t = (1/100) \times T$. According to the rule in (3.8), the fractional uncertainty in $q = Bx$ is the sum of those in B and x. Since $\delta B = 0$,

$$\frac{\delta q}{|q|} = \frac{\delta x}{|x|}.$$

Multiplying through by $|q| = |Bx|$, we find that $\delta q = |B|\,\delta x$, and we have the following useful rule.

Measured Quantity Times Exact Number

If the quantity x is measured with uncertainty δx and is used to compute the product

$$q = Bx$$

where B has no uncertainty, then the uncertainty in q is just $|B|$ times that in x,

$$\delta q = |B|\,\delta x. \tag{3.9}$$

This rule is especially useful when one has to measure something that is inconveniently small, but is available many times over, such as the thickness of a sheet of paper or the time for a revolution of a rapidly spinning wheel. For example, if we measure the thickness T of 100 sheets of paper and get the answer

$$\text{thickness of 100 sheets} = T = 1.3 \pm .1 \text{ inches,}$$

then it immediately follows that the thickness t of a single sheet is

$$\begin{aligned} \text{thickness of one sheet} = t &= \frac{1}{100} \times T \\ &= .013 \pm .001 \text{ inches.} \end{aligned}$$

Notice how this technique (measuring the thickness of several identical sheets and dividing by their number) makes easily possible a measurement that would otherwise require quite sophisticated equipment, and also gives a remarkably small uncertainty. One must, of course, be sure that all sheets are equally thick.

Powers

The second special case of the rule in (3.8) concerns the evaluation of a power of some measured quantity. For example, we might measure the speed v of some body and then, in order to find the kinetic energy $\frac{1}{2}mv^2$, calculate the square v^2. Since v^2 is just $v \times v$, it follows from (3.8) that the fractional uncertainty in v^2 is *twice* the fractional uncertainty in v. More generally, it is clear from (3.8) that the general rule for any power is as follows.

Uncertainty in a Power

If the quantity x is measured with uncertainty δx, and the measured value used to compute the power

$$q = x^n,$$

then the fractional uncertainty in q is n times that in x,

$$\frac{\delta q}{|q|} = n \frac{\delta x}{|x|}. \tag{3.10}$$

Our derivation of this rule required that n be a positive integer. In fact, however, the rule generalizes to include *any* exponent n. See (3.26) below.

Example

Suppose a student measures g, the acceleration of gravity, by measuring the time t for a stone to fall from a height h above the ground. After making several timings, he concludes that

$$t = 1.6 \pm .1 \text{ sec},$$

and he measures the height h as

$$h = 46.2 \pm .3 \text{ ft}.$$

Since h is given by the well-known formula $h = \frac{1}{2}gt^2$, he now calculates g as

$$\begin{aligned} g &= \frac{2h}{t^2} \\ &= \frac{2 \times 46.2 \text{ ft}}{(1.6 \text{ sec})^2} \\ &= 36.1 \text{ ft/sec}^2. \end{aligned}$$

The uncertainty in this answer can be found by using the rules just developed. To this end we need to know the fractional uncertainties in each of the factors in the expression $g = 2h/t^2$ used to calculate g. The factor 2 has no uncertainty. The fractional uncertainties in h and t are

$$\frac{\delta h}{h} = \frac{.3}{46.2} = .7\%$$

and

$$\frac{\delta t}{t} = \frac{.1}{1.6} = 6.3\%.$$

According to the rule in (3.10), the fractional uncertainty of t^2 is twice that of t. Therefore, applying the rule in (3.8) for products and quotients to the formula $g = 2h/t^2$, we find the fractional uncertainty

$$\begin{aligned} \frac{\delta g}{g} &= \frac{\delta h}{h} + 2\frac{\delta t}{t} \\ &= .7\% + 2 \times (6.3\%) = 13.3\%, \end{aligned} \tag{3.11}$$

and hence the uncertainty

$$\delta g = (36.1 \text{ ft/sec}^2) \times \frac{13.3}{100} = 4.80 \text{ ft/sec}^2.$$

Thus our student's final answer (properly rounded) is

$$g = 36 \pm 5 \text{ ft/sec}^2.$$

This example illustrates how simple the estimation of uncertainties can often be. It also illustrates how error analysis tells one not only the size of uncertainties, but also what must be done to reduce them. In this example, it is clear from (3.11) that the largest contribution comes from the measurement of the time. If we want a more precise value of g, then it is the measurement of t which must be improved; any attempt to improve the measurement of h will be so much wasted effort.

3.3. *Independent Uncertainties in a Sum*

The rules that we have found so far can be quickly summarized: when measured quantities are added or subtracted, the *uncertainties add*; when measured quantities are multiplied or divided, the *fractional uncertainties add.* In this and the next section, we will discuss how, under certain conditions, the uncertainties calculated by using these rules may be unnecessarily large. Specifically, we will see that if the original uncertainties are *independent* and *random*, then a more realistic (and smaller) estimate of the final uncertainty is given by similar rules, in which the uncertainties (or fractional uncertainties) are *added in quadrature* (a procedure we will define shortly).

Let us first consider computing the sum, $q = x + y$, of two numbers x and y that have been measured in the standard form

$$(\text{measured value of } x) = x_{\text{best}} \pm \delta x,$$

with a similar expression for y. The argument used in the last section was as follows. First, the best estimate for $q = x + y$ is obviously $q_{\text{best}} = x_{\text{best}} + y_{\text{best}}$. Second, since the highest probable values for x and y are $x_{\text{best}} + \delta x$ and $y_{\text{best}} + \delta y$, the highest probable value for q is

$$x_{\text{best}} + y_{\text{best}} + \delta x + \delta y. \tag{3.12}$$

Similarly, the lowest probable value of q is

$$x_{\text{best}} + y_{\text{best}} - \delta x - \delta y.$$

Therefore, we concluded, the value of q probably lies between these two numbers, and the uncertainty in q is

$$\delta q \approx \delta x + \delta y.$$

To see why this formula is likely to overestimate δq, let us consider how it could happen that the actual value of q is equal to the highest extreme (3.12). Obviously, this happens if we have underestimated x by the full amount δx *and* underestimated y by the full δy. And obviously it is fairly unlikely that this would happen. If x and y are measured independently and our errors are random in nature, there is a 50 percent chance that an *underestimate* of x will be accompanied by an *overestimate* of y, or *vice versa.* Clearly, then, the probability that we will underestimate both x and y by the full amounts δx and δy is fairly small. Therefore the value $\delta q \approx \delta x + \delta y$ overestimates our probable error.

Just what constitutes a better estimate of δq? That depends on precisely what we mean by uncertainties (i.e., what we mean by the statement that q is "probably" somewhere between $q_{\text{best}} - \delta q$ and $q_{\text{best}} + \delta q$). It also depends on what are the statistical laws governing our errors in measurement. In Chapter 5 we will discuss the normal, or Gauss, distribution, which describes measurements subject to random uncertainties. We will see that if the measurements of x and y are made independently, and are both governed by the normal distribution, then the uncertainty in $q = x + y$ is given by

$$\delta q = \sqrt{(\delta x)^2 + (\delta y)^2}. \tag{3.13}$$

When we combine two numbers by squaring them, adding the squares, and then taking the square root, as in (3.13), the numbers are said to be *added in quadrature*. Thus the rule embodied in (3.13) can be stated as follows: if the measurements of x and y are independent and subject only to random uncertainties, then the uncertainty δq in the calculated value of $q = x + y$ is the *sum in quadrature* or *quadratic sum* of the uncertainties δx and δy.

It is important to compare the new expression (3.13) for the uncertainty in $q = x + y$ with our old expression,

$$\delta q \approx \delta x + \delta y. \tag{3.14}$$

First, the new expression (3.13) is always smaller than the old (3.14), as we can see from a simple geometrical argument. For any two positive numbers a and b, the numbers a, b, and $\sqrt{a^2 + b^2}$ are the three sides of a right-angled triangle (Figure 3.2). Since the length of any side of a triangle is always less than the sum of the other two sides, it follows that $\sqrt{a^2 + b^2} < a + b$ and hence that (3.13) is always less than (3.14).

Since expression (3.13) for the uncertainty in $q = x + y$ is always smaller than (3.14), one should always use (3.13) *when* it is applicable. However, it is *not* always applicable. Expression (3.13) reflects the possibility that an overestimate in x can be offset by an underestimate in y or vice versa. It is easy to think of measurements where this is not possible.

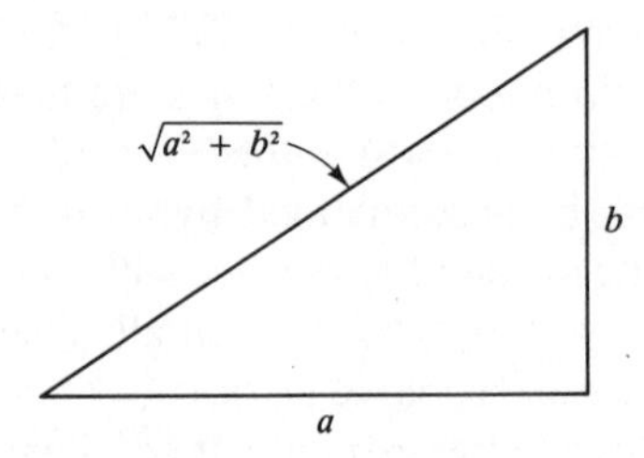

Figure 3.2. Since any side of a triangle is less than the sum of the other two sides, it is always true that $\sqrt{a^2 + b^2} < a + b$.

Suppose, for example, that $q = x + y$ is the sum of two lengths x and y measured with the same steel tape. Suppose further that the main source of uncertainty is our fear that the tape was designed for use at a temperature different from the present temperature. If we don't know this temperature (and don't have a reliable tape for comparison), then we have to recognize that our tape may be longer or shorter than its calibrated length, and hence may be reading under or over the correct length. This uncertainty can be easily allowed for.[3] However, the point here is that if the tape is too long, then we *underestimate both* x and y; and if the tape is too short, we *overestimate both* x and y. Thus there is no possibility for the cancellations that justified using the sum in quadrature to compute the uncertainty in $q = x + y$.

We will prove later (in Chapter 9) that, whether or not our errors are independent and random, the uncertainty in $q = x + y$ is *certainly no larger* than the simple sum $\delta x + \delta y$:

$$\delta q \leqslant \delta x + \delta y. \tag{3.15}$$

That is, our old expression (3.14) for δq is actually an *upper bound* which holds in all cases. If we have any reason to suspect the errors in x and y are *not* independent and random (as in the example of the steel tape measure), then we are not justified in using the quadratic sum (3.13) for δq. On the other hand, the bound (3.15) guarantees that δq is certainly no worse than $\delta x + \delta y$, and our safest course is to use the old rule

$$\delta q \approx \delta x + \delta y.$$

Often it actually makes little difference whether one adds uncertainties in quadrature or directly. For example, suppose that x and y are lengths both measured with uncertainties $\delta x = \delta y = 2$ mm. If we are sure that these uncertainties are independent and random, then we would estimate the error in $x + y$ to be the sum in quadrature,

$$\sqrt{(\delta x)^2 + (\delta y)^2} = \sqrt{4 + 4}\text{ mm} = 2.8\text{ mm} \approx 3\text{ mm},$$

but if we suspect that the uncertainties may not be independent, then we would have to use the ordinary sum,

$$\delta x + \delta y \approx (2 + 2)\text{ mm} = 4\text{ mm}.$$

[3] Suppose, for example, that the tape has a coefficient of expansion $\alpha = 10^{-5}$ per degree, and that we decide that the difference between its calibration temperature and the present temperature is unlikely to be more than 10 degrees. The tape is then unlikely to be more than 10^{-4}, or .01 percent, away from its correct length, and our uncertainty is therefore .01 percent.

In many experiments, the estimation of uncertainties is so crude that the difference between these two answers (3 mm and 4 mm) is unimportant. On the other hand, sometimes the sum in quadrature is significantly smaller than the ordinary sum. Also, rather surprisingly, the sum in quadrature is sometimes easier to compute than the ordinary sum. We will see examples of these effects in the next section.

3.4. More About Independent Uncertainties

In the last section we discussed how independent random uncertainties in two quantities x and y propagate to cause an uncertainty in the sum $x + y$. We saw that for this type of uncertainty the two errors should be added in quadrature. One can naturally consider the corresponding problem for differences, products, and quotients. As we will prove later, one can show that in all cases our previous rules, (3.4) and (3.8), are modified only in that the sums of errors (or fractional errors) are replaced by quadratic sums. Further, we shall prove that the old expressions (3.4) and (3.8) are, in fact, upper bounds, which always hold, whether or not the uncertainties are independent and random. Thus the final versions of our two main rules are as follows:

Uncertainty in Sums and Differences

Suppose that $x, \ldots, w$ are measured with uncertainties $\delta x, \ldots, \delta w$, and the measured values used to compute

$$q = x + \cdots + z - (u + \cdots + w).$$

If the uncertainties in $x, \ldots, w$ are known to be *independent and random*, then the uncertainty in q is the quadratic sum

$$\delta q = \sqrt{(\delta x)^2 + \cdots + (\delta z)^2 + (\delta u)^2 + \cdots + (\delta w)^2} \tag{3.16}$$

of the original uncertainties. In any case, δq is never larger than their ordinary sum,

$$\delta q \leqslant \delta x + \cdots + \delta z + \delta u + \cdots + \delta w. \tag{3.17}$$

and

Uncertainties in Products and Quotients

Suppose that $x, \ldots, w$ are measured with uncertainties $\delta x, \ldots, \delta w$, and the measured values used to compute

$$q = \frac{x \times \cdots \times z}{u \times \cdots \times w}.$$

If the uncertainties in $x, \ldots, w$ are *independent and random*, then the fractional uncertainty in q is the sum in quadrature of the original fractional uncertainties,

$$\frac{\delta q}{|q|} = \sqrt{\left(\frac{\delta x}{x}\right)^2 + \cdots + \left(\frac{\delta z}{z}\right)^2 + \left(\frac{\delta u}{u}\right)^2 + \cdots + \left(\frac{\delta w}{w}\right)^2}. \tag{3.18}$$

In any case, it is never larger than their ordinary sum,

$$\frac{\delta q}{|q|} \leqslant \frac{\delta x}{|x|} + \cdots + \frac{\delta z}{|z|} + \frac{\delta u}{|u|} + \cdots + \frac{\delta w}{|w|}. \tag{3.19}$$

Notice that we have not yet justified the use of addition in quadrature for independent random uncertainties. We have only argued that when the various uncertainties are independent and random, there is a good chance of partial cancellations of errors, and that the resulting uncertainty (or fractional uncertainty) should be smaller than the simple sum of the original uncertainties (or fractional uncertainties); the sum in quadrature does have this property. We will give a proper justification of its use in Chapter 5. The bounds (3.17) and (3.19) will be proved in Chapter 9.

Example

As we have already discussed, sometimes there is no significant difference between uncertainties computed by addition in quadrature and those computed by straight addition. On the other hand, there often is a significant difference, and—surprisingly enough—the sum in quadrature is often much simpler to compute. To see how this can happen, consider the following example.

Suppose that we wish to find the efficiency of a D.C. electric motor by using it to lift a mass m through a height h. The work accomplished is mgh, and the electric energy delivered to the motor is VIt, where V is the applied voltage, I the current, and t the time for which the motor runs. The efficiency is then

$$\text{efficiency, } e = \frac{\text{work done by motor}}{\text{energy delivered to motor}} = \frac{mgh}{VIt}.$$

Let us suppose that m, h, V, and I can all be measured with 1 percent accuracy,

$$(\text{fractional uncertainty for } m, h, V, \text{ and } I) = 1\%,$$

and that the time t has an uncertainty of 5 percent,

$$(\text{fractional uncertainty for } t) = 5\%.$$

(Of course, g is known with negligible uncertainty.) If we now compute the efficiency e, then, according to our old rule ("fractional errors add"), we have an uncertainty

$$\frac{\delta e}{e} \approx \frac{\delta m}{m} + \frac{\delta h}{h} + \frac{\delta V}{V} + \frac{\delta I}{I} + \frac{\delta t}{t}$$
$$= (1 + 1 + 1 + 1 + 5)\% = 9\%.$$

On the other hand, if we are confident that the various uncertainties are independent and random, then we can compute $\delta e/e$ by the quadratic sum to give

$$\frac{\delta e}{e} = \sqrt{\left(\frac{\delta m}{m}\right)^2 + \left(\frac{\delta h}{h}\right)^2 + \left(\frac{\delta V}{V}\right)^2 + \left(\frac{\delta I}{I}\right)^2 + \left(\frac{\delta t}{t}\right)^2}$$
$$= \sqrt{1^2 + 1^2 + 1^2 + 1^2 + 5^2}\,\%$$
$$= \sqrt{29}\,\% \approx 5\%.$$

Clearly, the quadratic sum leads to a significantly smaller estimate for δe. Furthermore, it will be seen that, to one significant figure, the uncertainties in m, h, V, and I *make no contribution at all* to the uncertainty in e computed in this way; that is, to one significant figure, we have found (in this example)

$$\frac{\delta e}{e} = \frac{\delta t}{t}.$$

This striking simplification is easily understood. When numbers are added in quadrature, they are squared first and then summed. The process of squaring greatly exaggerates the importance of the larger numbers. Thus, if one number is 5 times any of the others (as in our example), then its square is 25 times that of the others, and we can usually neglect the others entirely.

This example illustrates how it is usually better, and often easier, to combine errors in quadrature. The example also illustrates what the type of problem is in which the errors *are* independent, and for which addition in quadrature is justified. (For the moment we take for granted that the errors are random. We will discuss this more difficult point in Chapter 4.) The five quantities measured (m, h, V, I, and t) are physically distinct quantities, with different units, and are measured by entirely different processes. It is almost inconceivable that the sources of error in any one quantity are correlated with those in any other. Therefore the errors can reasonably be treated as independent and combined in quadrature.

3.5. *Arbitrary Functions of One Variable*

We now know how uncertainties, both independent and otherwise, propagate through sums, differences, products, and quotients. However, many calculations require more complicated operations, like computation of a sine, cosine, or square root, and we will need to know how uncertainties propagate in these cases.

As an example, we could imagine finding the refractive index n of glass by measuring the critical angle θ. It is known from elementary optics that $n = 1/\sin\theta$. If we can measure the angle θ, it is then easy to calculate the refractive index n. But we must then decide what uncertainty δn in $n = 1/\sin\theta$ results from the uncertainty $\delta\theta$ in our measurement of θ.

More generally, suppose we have measured a quantity x in the standard form $x_{\text{best}} \pm \delta x$, and wish to calculate some known function $q(x)$, such as $q(x) = 1/\sin x$ or $q(x) = \sqrt{x}$. One simple way to think about this calculation is to draw a graph of $q(x)$ as in Figure 3.3. The best estimate for

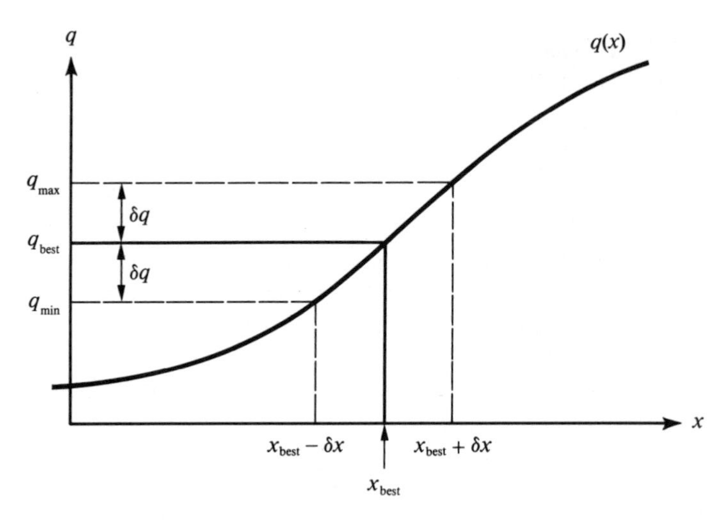

Figure 3.3. Graph of $q(x)$ vs. x. If x is measured as $x_{best} \pm \delta x$, then the best estimate for $q(x)$ is $q_{best} = q(x_{best})$. The largest and smallest probable values of $q(x)$ correspond to the values $x_{best} \pm \delta x$ of x.

$q(x)$ is, of course, $q_{best} = q(x_{best})$, and the values x_{best} and q_{best} are shown connected by the heavy lines in Figure 3.3.

To decide on the uncertainty δq, we employ the usual argument. The largest probable value of x is $x_{best} + \delta x$; using the graph, we can immediately find the largest probable value of q, which is shown as q_{max}. Similarly, we can draw in the smallest probable value, q_{min}, as shown. If the uncertainty δx is small (as we always suppose it is), then the section of graph involved in this construction is approximately straight, and it is easily seen that q_{max} and q_{min} are equally spaced on either side of q_{best}. The uncertainty δq can then be taken from the graph as either of the lengths shown, and we have found the value of q in the standard form $q_{best} \pm \delta q$.

Occasionally one actually calculates uncertainties from a graph as just described. (See Problem 3.10 for an example.) Usually, however, the function $q(x)$ is known explicitly—$q(x) = \sin x$ or $q(x) = \sqrt{x}$, for example—and the uncertainty δq can be calculated analytically. It is clear from Figure 3.3 that

$$\delta q = q(x_{best} + \delta x) - q(x_{best}). \tag{3.20}$$

Now, a fundamental approximation of calculus asserts that, for any function $q(x)$ and any sufficiently small increment u,

$$q(x + u) - q(x) = \frac{dq}{dx} u.$$

Thus, provided the uncertainty δx is small (as we always assume it is), we can rewrite the difference in (3.20) to give

$$\delta q = \frac{dq}{dx} \delta x. \tag{3.21}$$

Thus, to find the uncertainty δq, we just calculate the derivative dq/dx and multiply by the uncertainty δx.

The rule (3.21) is not quite in its final form. It was derived for a function, like that of Figure 3.3, whose slope is positive. In Figure 3.4 is shown a function with negative slope. Here the maximum probable value q_{max} obviously corresponds to the minimum value, $x_{best} - \delta x$, of x, so that

$$\delta q = -\frac{dq}{dx} \delta x. \tag{3.22}$$

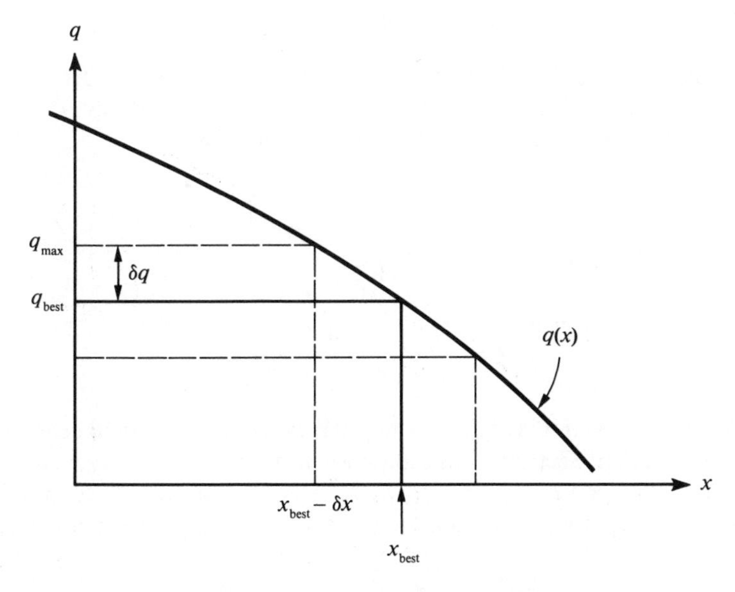

Figure 3.4. If the slope of $q(x)$ is negative, then the maximum probable value of q corresponds to the minimum value of x, and vice versa.

Since dq/dx is negative, we can write $-dq/dx$ as $|dq/dx|$, and we have the following general rule.

Uncertainty in Any Function of One Variable

If x is measured with uncertainty δx and is used to calculate the function $q(x)$, then the uncertainty δq is

$$\delta q = \left|\frac{dq}{dx}\right| \delta x. \tag{3.23}$$

As a simple application of this rule, suppose that we have measured an angle θ as

$$\theta = 20 \pm 3 \text{ deg},$$

and that we wish to find $\cos\theta$. Our best estimate of $\cos\theta$ is, of course, $\cos 20^\circ = .94$, and according to (3.23) the uncertainty is

$$\begin{aligned} \delta(\cos\theta) &= \left|\frac{d\cos\theta}{d\theta}\right| \delta\theta \\ &= |\sin\theta|\,\delta\theta \text{ (in rad).} \end{aligned} \tag{3.24}$$

We have indicated that $\delta\theta$ must be expressed in radians, because the derivative of $\cos\theta$ is $-\sin\theta$ only if θ is expressed in radians. Therefore we rewrite $\delta\theta = 3^\circ$ as $\delta\theta = .05$ rad; then (3.24) gives

$$\begin{aligned} \delta(\cos\theta) &= (\sin 20^\circ) \times .05 \\ &= 0.34 \times .05 \\ &= 0.02. \end{aligned}$$

Thus our final answer is

$$\cos\theta = 0.94 \pm 0.02.$$

As a second example of the rule in (3.23) we can rederive (and generalize) a result found in Section 3.2. Suppose we measure the quantity x and then calculate the power, $q(x) = x^n$ (where n is any known, fixed number, positive or negative). According to (3.23) the resulting uncertainty in q is

$$\delta q = \left|\frac{dq}{dx}\right| \delta x = |nx^{n-1}|\,\delta x.$$

If we divide both sides of this equation by $|q| = |x^n|$, we find that

$$\frac{\delta q}{|q|} = |n| \frac{\delta x}{|x|}; \tag{3.25}$$

that is, the fractional uncertainty in $q = x^n$ is $|n|$ times that in x. This is just the rule (3.10) found earlier. However, the result here is more general, since n can now be any number. For example, if $n = \frac{1}{2}$, then $q = \sqrt{x}$, and

$$\frac{\delta q}{|q|} = \frac{1}{2} \frac{\delta x}{|x|};$$

that is, the fractional uncertainty in $\sqrt{x}$ is *half* that in x itself. Similarly, the fractional uncertainty in $1/x = x^{-1}$ is the same as that in x itself.

The result (3.25) is just a special case of the rule in (3.23). However, it is sufficiently important to deserve separate statement as the following general rule.

Uncertainty in a Power

If x is measured with uncertainty δx and is used to calculate the power $q = x^n$ (where n is a fixed, known number), then the fractional uncertainty in q is $|n|$ times that in x,

$$\frac{\delta q}{|q|} = |n| \frac{\delta x}{|x|}. \tag{3.26}$$

3.6. Propagation Step by Step

We now have enough tools to handle almost any problem in the propagation of errors. Any calculation can be broken down into a sequence of steps, each involving just one of the following types of operation: (1) sums and differences; (2) products and quotients; and (3) computation of a function of one variable, like x^n, $\sin x$, e^x, or $\ln x$. For example, we could calculate

$$q = x(y - z \sin u) \tag{3.27}$$

from the measured quantities x, y, z, and u in the following steps: compute the *function* $\sin u$, then the *product* of z and $\sin u$, next the *difference* of y and $z \sin u$, and finally the *product* of x and $(y - z \sin u)$.

We know how uncertainties propagate through each of these separate operations. Thus, provided the various quantities involved are independent, we can calculate the uncertainty in the final answer by proceeding in steps from the uncertainties in the original measurement.[4] For example, if the quantities x, y, z, and u in (3.27) have been measured with corresponding uncertainties $\delta x, \ldots, \delta u$, then we could calculate the uncertainty in q as follows. First find the uncertainty in the function $\sin u$; knowing this, find the uncertainty in the product $z \sin u$, and then that in the difference $y - z \sin u$; finally, find the uncertainty in the complete product (3.27).

Before we discuss some examples of this step-by-step calculation of errors, let us emphasize two general points. First, since uncertainties in sums or differences involve absolute uncertainties (like δx), whereas those in products or quotients involve fractional uncertainties (like $\delta x/|x|$), our calculations will require some facility in passing from absolute to fractional uncertainties and vice versa, as we will see.

Second, an important simplifying feature of all these calculations is that (as we have repeatedly emphasized) uncertainties are seldom needed to more than one significant figure. Hence much of the calculation can be done very rapidly in one's head, and many smaller uncertainties can be completely neglected. In a typical experiment involving several trials, one may need to do a careful calculation on paper of all error propagations for the first trial. Once this has been done, it is often easy to see that all trials are sufficiently similar that no further calculation is needed or, at worst, that for subsequent trials the calculations of the first trial can be modified in one's head.

3.7. *Examples*

In this and the next section we detail three examples of the type of calculation encountered in introductory laboratories. None of these examples is especially complicated; and, in fact, few real problems are much more complicated than the ones treated here.

[4] We will discuss in Section 3.9 why this step-by-step procedure is sometimes unsatisfactory when the various quantities are not independent, as with a function like $q = x(y - x \sin y)$, where x and y appear twice over. Here a step-by-step calculation of the uncertainty δq may sometimes overestimate δq.

Measurement of g with a Simple Pendulum

As a first example, suppose that we measure g, the acceleration of gravity, using a simple pendulum. The period of such a pendulum is well-known to be $T = 2\pi\sqrt{l/g}$, where l is the length of the pendulum. Thus if l and T are measured, we can find g as

$$g = 4\pi^2 l/T^2. \tag{3.28}$$

This gives g as the product or quotient of three factors, $4\pi^2$, l, and T^2. If the various uncertainties are independent and random, the fractional uncertainty in our answer is just the quadratic sum of the fractional uncertainties in these factors. The factor $4\pi^2$ has no uncertainty, and the fractional uncertainty in T^2 is twice that in T:

$$\frac{\delta(T^2)}{T^2} = 2\frac{\delta T}{T}.$$

Thus the fractional uncertainty in our answer for g will be

$$\frac{\delta g}{g} = \sqrt{\left(\frac{\delta l}{l}\right)^2 + \left(2\frac{\delta T}{T}\right)^2}. \tag{3.29}$$

Suppose we measure the period T for one value of the length l and get the results[5]

$$l = 92.95 \pm .1 \text{ cm},$$
$$T = 1.936 \pm .004 \text{ sec}.$$

Our best estimate for g is easily found from (3.28) as

$$g_{\text{best}} = \frac{4\pi^2 \times (92.95 \text{ cm})}{(1.936 \text{ sec})^2} = 979 \text{ cm/sec}^2.$$

To find our uncertainty in g using (3.29), we need the fractional uncertainties in l and T. These are easily calculated (in the head) as

$$\frac{\delta l}{l} = 0.1\% \quad \text{and} \quad \frac{\delta T}{T} = 0.2\%.$$

[5] Although at first sight an uncertainty $\delta T = .004$ sec may seem unrealistically small, one can easily achieve it by timing several oscillations. If one can measure with an accuracy of .1 sec, as is certainly possible with a stopwatch, then by timing 25 oscillations one will find T within .004 sec.

Substituting into (3.29), we find

$$\frac{\delta g}{g} = \sqrt{(0.1)^2 + (2 \times 0.2)^2}\,\% = 0.4\%;$$

whence

$$\begin{aligned}\delta g &= 0.004 \times 979 \text{ cm/sec}^2 \\ &= 4 \text{ cm/sec}^2.\end{aligned}$$

Thus our final answer, based on these measurements, is

$$g = 979 \pm 4 \text{ cm/sec}^2.$$

If this experiment is now repeated (as most such experiments should be) with different values for the parameters, it will not be necessary to repeat the uncertainty calculations in complete detail. With a little thought, one can easily record the various values of l, T, and g *and* the corresponding uncertainty calculations all in a single tabulation (see Problem 3.13).

Refractive Index Using Snell's Law

If a ray of light passes from air into glass, then the angles of incidence (i) and refraction (r) are defined as in Figure 3.5 and are related by Snell's law, $\sin i = n \sin r$, where n is the refractive index of the glass. Thus, if one measures the angles i and r, one can calculate the refractive index n as

$$n = \sin i/\sin r. \tag{3.30}$$

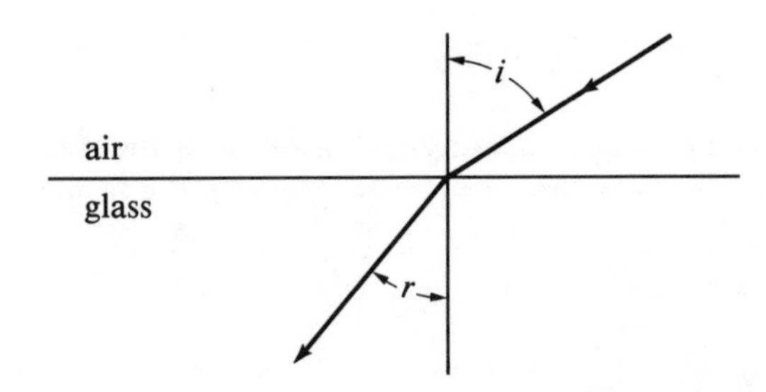

Figure 3.5. The angles of incidence, i, and refraction, r, when a ray of light passes from air into glass.

The uncertainty in this answer is easily calculated. Since n is the quotient of $\sin i$ and $\sin r$, the fractional uncertainty in n is the quadratic sum,

$$\frac{\delta n}{n} = \sqrt{\left(\frac{\delta \sin i}{\sin i}\right)^2 + \left(\frac{\delta \sin r}{\sin r}\right)^2}, \tag{3.31}$$

of those in $\sin i$ and $\sin r$. To find the fractional uncertainty in the sine of any angle θ, we note that

$$\begin{aligned}\delta \sin \theta &= \left|\frac{d \sin \theta}{d\theta}\right| \delta\theta \\ &= |\cos \theta|\, \delta\theta \text{ (in rad).}\end{aligned}$$

Thus the fractional uncertainty is

$$\frac{\delta \sin \theta}{|\sin \theta|} = |\cot \theta|\, \delta\theta \text{ (in rad).} \tag{3.32}$$

Suppose we now measure the angle r for a couple of values of i, and get the results shown in the first two columns of Table 3.1 (with all measurements judged to be uncertain by ± 1 degree, or 0.02 radians). The calculation of $n = \sin i/\sin r$ is easily carried out as shown in the next three columns of Table 3.1. The uncertainty in n can then be found as in the last three columns; the fractional uncertainties in $\sin i$ and $\sin r$ are calculated using (3.32), and finally that in n is found using (3.31).

Table 3.1. Finding the refractive index.

i (deg) all ± 1	r (deg) all ± 1	$\sin i$	$\sin r$	n	$\frac{\delta \sin i}{\lvert\sin i\rvert}$	$\frac{\delta \sin r}{\lvert\sin r\rvert}$	$\frac{\delta n}{n}$
20	13	.342	.225	1.52	5 %	8 %	9 %
40	23.5	.643	.399	1.61	2 %	4 %	5 %

Before making a series of measurements like the two shown in Table 3.1, think carefully how best to record the data and calculations. A tidy display like that in Table 3.1 makes the recording of data easier and reduces the danger of mistakes in calculation. It is also easier for the reader to follow and to check.

3.8. A More Complicated Example

The two examples just given are typical of many experiments in the introductory physics laboratory. However, a few experiments require more complicated calculations. As an example of such an experiment, we discuss here the measurement of the acceleration of a cart rolling down a slope.[6]

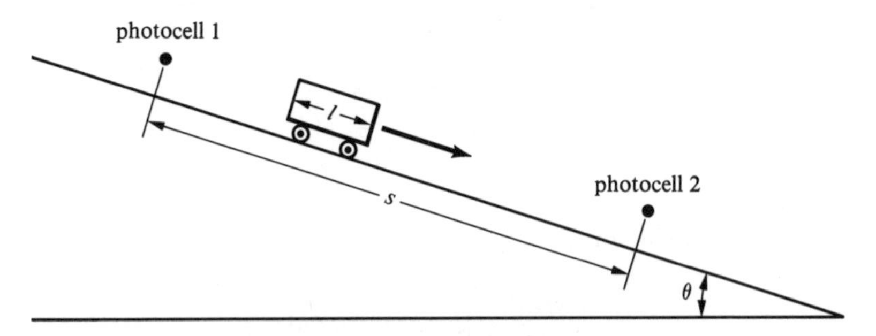

Figure 3.6. A cart rolls down an incline of slope θ. Each photocell is connected to a timer to measure the time for the cart to pass it.

Acceleration of a Cart Down a Slope

Let us consider a cart rolling down an incline of slope θ as in Figure 3.6. The expected acceleration is $g \sin \theta$ and, if we measure θ, we can easily calculate the expected acceleration and its uncertainty (Problem 3.15). We can measure the actual acceleration, a, by timing the cart past two photocells as shown, each connected to a timer. If the cart has length l and takes time t_1 to pass the first photocell, its speed there is $v_1 = l/t_1$. In the same way, $v_2 = l/t_2$. (Strictly speaking, these speeds are the cart's *average* speeds while passing the two photocells. However, so long as l is small, the difference between the average and instantaneous speeds is unimportant.) If the distance between the photocells is s, then the well-known formula $v_2{}^2 = v_1{}^2 + 2as$ implies that

$$a = \frac{v_2{}^2 - v_1{}^2}{2s}$$

$$= \left(\frac{l^2}{2s}\right)\left(\frac{1}{t_2{}^2} - \frac{1}{t_1{}^2}\right). \qquad (3.33)$$

[6] The reader who wishes could omit this section without loss of continuity, or might return to study the section in connection with Problem 3.15.

Using this formula and the measured values of l, s, t_1, and t_2, one can easily find the observed acceleration and its uncertainty.

A set of data for this experiment (where the numbers in parentheses are the corresponding percentage uncertainties, as you can easily check) was as follows:

$$\begin{aligned} l &= 5.0 \pm .05 \text{ cm} && (1\%) \\ s &= 100.0 \pm .2 \text{ cm} && (.2\%) \\ t_1 &= .054 \pm .001 \text{ sec} && (2\%) \\ t_2 &= .031 \pm .001 \text{ sec} && (3\%). \end{aligned} \tag{3.34}$$

From these we can immediately calculate the first factor in (3.33) as $l^2/2s = .125$ cm. Since the fractional uncertainties in l and s are 1 and .2 percent, that in $l^2/2s$ is

$$\sqrt{(2 \times 1)^2 + (.2)^2}\,\% = 2\%.$$

(Note how the uncertainty in s makes no appreciable contribution and could have been ignored.) Therefore

$$l^2/2s = .125 \text{ cm} \pm 2\%. \tag{3.35}$$

To calculate the second factor in (3.33) and its uncertainty, we proceed in steps. Since the fractional uncertainty in t_1 is 2 percent, that in $1/t_1{}^2$ is 4 percent. Thus, since $t_1 = .054$ sec,

$$1/t_1{}^2 = 343 \pm 14 \text{ sec}^{-2}.$$

In the same way, the fractional uncertainty in $1/t_2{}^2$ is 6 percent and

$$1/t_2{}^2 = 1041 \pm 62 \text{ sec}^{-2}.$$

Subtracting these (and combining the errors in quadrature), we find

$$\frac{1}{t_2{}^2} - \frac{1}{t_1{}^2} = 698 \pm 64 \text{ sec}^{-2} \ (9\%). \tag{3.36}$$

Finally, according to (3.33), the required acceleration is the product of (3.35) and (3.36). Multiplying these together (and combining the fractional uncertainties in quadrature), we obtain

$$a = (.125\ \text{cm} \pm 2\%) \times (698\ \text{sec}^{-2} \pm 9\%)$$
$$= 87.3\ \text{cm/sec}^2 \pm 9\%$$

or

$$a = 87 \pm 8\ \text{cm/sec}^2. \qquad (3.37)$$

This answer could now be compared with the expected acceleration $g \sin \theta$, if it had been calculated.

When the calculations leading to (3.37) are studied carefully, several interesting features emerge. First, the 2 percent uncertainty in the factor $l^2/2s$ is completely swamped by the 9 percent uncertainty in $(1/t_2^2) - (1/t_1^2)$. If further calculations are needed for subsequent trials, the uncertainties in l and s can therefore be ignored (so long as one checks quickly to see that they are still just as unimportant).

Another important feature of our calculation is the way in which the 2 and 3 percent uncertainties in t_1 and t_2 grow when we evaluate $1/t_1^2$, $1/t_2^2$, and the difference $(1/t_2^2) - (1/t_1^2)$, so that the final uncertainty is 9 percent. This growth results partly from taking squares and partly from taking the difference of large numbers. We could imagine extending the experiment to check the constancy of a by giving the cart an initial push, so that the speeds v_1 and v_2 are both larger. If we did this, the times t_1 and t_2 would get smaller, and the effects just described would get worse (see Problem 3.15).

3.9. General Formula for Error Propagation[7]

So far we have established three main rules for the propagation of errors: that for sums and differences; that for products and quotients; and that for arbitrary functions of one variable. In the last three sections we have seen how the computation of a complicated function can often be broken into steps, and the uncertainty in the function computed step by step using our three simple rules.

In this final section we give a single general formula from which all three of these rules can be derived, and with which any problem in error propagation can be solved. Although this formula is usually rather cumbersome to use, it is useful theoretically. Furthermore, there are some

[7] The reader can postpone reading this section without serious loss of continuity. The material covered here is not used again until Section 5.6.

problems where, instead of calculating the uncertainty in steps as in the last three sections, it is better to do the calculation in one step by means of the general formula.

To illustrate the kind of problem for which the one-step calculation is preferable, suppose that we measure three quantities x, y, z, and have to compute a function like

$$q = \frac{x + y}{x + z} \tag{3.38}$$

in which a variable appears more than once (x in this case). If we were to calculate the uncertainty δq in steps, then we would first compute the uncertainties in the two sums $x + y$ and $x + z$, and then that in their quotient. Proceeding in this way, we would completely miss the possibility that errors in the numerator due to errors in x may, to some extent, cancel errors in the denominator due to errors in x. To understand how this can happen, suppose that x, y, z are all positive numbers, and consider what happens if our measurement of x is subject to error. If we *overestimate* x, then we *overestimate both* $x + y$ and $x + z$, and (to a large extent) these overestimates cancel one another when we calculate $(x + y)/(x + z)$. Similarly, an *underestimate* of x leads to *underestimates* of *both* $x + y$ and $x + z$, which again cancel when we form the quotient. In either case, an error in x is substantially canceled out of the quotient $(x + y)/(x + z)$, and our step-by-step calculation completely misses these cancellations.

Whenever a function involves the same quantity more than once, as in (3.38), some of the errors may cancel themselves (an effect sometimes called *compensating errors*). If this is possible, then a step-by-step calculation of the uncertainty may overestimate the final uncertainty. The only way to avoid this is to calculate the uncertainty in one step by using the method that we will now develop.[8]

Let us suppose at first that we measure two quantities x and y, and then calculate some function $q = q(x, y)$. This function could be as simple as $q = x + y$ or something more complicated, like $q = (x^3 + y) \sin(xy)$. For a function $q(x)$ of a *single* variable, we argued that if the best estimate for x is the number x_{best}, then the best estimate for $q(x)$ is $q(x_{\text{best}})$. Next we argued that the extreme (i.e., largest and smallest) probable values of x are $x_{\text{best}} \pm \delta x$, and that the corresponding extreme values of q are therefore

$$q(x_{\text{best}} \pm \delta x). \tag{3.39}$$

[8] Sometimes a function that involves a variable more than once can be rewritten in a different form that does not. For example, $q = xy - xz$ can be rewritten as $q = x(y - z)$. In the second form the uncertainty δq can be calculated in steps without any danger of overestimation.

Finally, we used the approximation

$$q(x + u) \approx q(x) + \frac{dq}{dx}u \tag{3.40}$$

(for any small increment u) to rewrite the extreme probable values (3.39) as

$$q(x_{\text{best}}) \pm \left|\frac{dq}{dx}\right| \delta x, \tag{3.41}$$

where the absolute value is to allow for the possibility that dq/dx may be negative. The result (3.41) means that $\delta q \approx |dq/dx|\, \delta x$.

When q is a function of two variables, $q(x, y)$, the argument is very similar. If x_{best} and y_{best} are the best estimates for x and y, then we expect the best estimate for q to be

$$q_{\text{best}} = q(x_{\text{best}}, y_{\text{best}})$$

in the usual way. In order to estimate the uncertainty in this result, we need to generalize the approximation (3.40) for a function of two variables. The required generalization is

$$q(x + u, y + v) \approx q(x, y) + \frac{\partial q}{\partial x}u + \frac{\partial q}{\partial y}v, \tag{3.42}$$

where u and v are any small increments in x and y, and $\partial q/\partial x$ and $\partial q/\partial y$ are the so-called *partial derivatives* of q with respect to x and y. That is, $\partial q/\partial x$ is the result of differentiating q with respect to x while treating y as fixed, and vice versa for $\partial q/\partial y$. (For further discussion of partial derivatives, see Problems 3.16 and 3.17.)

The extreme probable values for x and y are $x_{\text{best}} \pm \delta x$ and $y_{\text{best}} \pm \delta y$. If we insert these into (3.42) and recall that $\partial q/\partial x$ and $\partial q/\partial y$ may be positive or negative, we find, for the extreme values of q,

$$q(x_{\text{best}}, y_{\text{best}}) \pm \left(\left|\frac{\partial q}{\partial x}\right| \delta x + \left|\frac{\partial q}{\partial y}\right| \delta y\right).$$

This means that the uncertainty in $q(x, y)$ is

$$\delta q \approx \left|\frac{\partial q}{\partial x}\right| \delta x + \left|\frac{\partial q}{\partial y}\right| \delta y. \tag{3.43}$$

Before we discuss various generalizations of this new rule, it is worth applying it to rederive some familiar cases. Suppose, for instance, that

$$q(x, y) = x + y; \tag{3.44}$$

that is, q is just the sum of x and y. The partial derivatives are both one,

$$\frac{\partial q}{\partial x} = \frac{\partial q}{\partial y} = 1, \tag{3.45}$$

and so, according to (3.43),

$$\delta q \approx \delta x + \delta y. \tag{3.46}$$

This is just the familiar rule that the uncertainty in $x + y$ is the sum of the uncertainties in x and y.

In much the same way, if q is the product $q = xy$, one can check that (3.43) implies the familiar rule that the fractional uncertainty in q is the sum of the fractional uncertainties in x and y (see Problem 3.18).

The rule in (3.43) can be generalized in various ways. The reader will not be surprised to learn that when the uncertainties δx and δy are independent and random, the sum (3.43) can be replaced by a sum in quadrature. If the function q depends on more than two variables, then we simply add an extra term for each extra variable. This brings us to the following general rule (whose proper justification will appear in Chapters 5 and 9).

Uncertainty in a Function of Several Variables

Suppose that $x, \ldots, z$ are measured with uncertainties $\delta x, \ldots, \delta z$, and the measured values used to compute the function $q(x, \ldots, z)$. If the uncertainties in $x, \ldots, z$ are independent and random, then the uncertainty in q is

$$\delta q = \sqrt{\left(\frac{\partial q}{\partial x}\delta x\right)^2 + \cdots + \left(\frac{\partial q}{\partial z}\delta z\right)^2}. \tag{3.47}$$

In any case, it is never larger than the ordinary sum

$$\delta q \leqslant \left|\frac{\partial q}{\partial x}\right|\delta x + \cdots + \left|\frac{\partial q}{\partial z}\right|\delta z. \tag{3.48}$$

Perhaps the most useful feature of this general rule is that we can derive from it all our earlier rules for error propagation (see Problem 3.18). The direct use of the general rule is usually fairly cumbersome in practice, and it is usually simpler, if possible, to proceed step by step using our previous simpler rules. However, if the function $q(x, \ldots, z)$ involves any variable more than once, then there may be compensating errors; if so, a step-by-step calculation may overestimate the final uncertainty, and it is better to calculate δq in one step using (3.47) or (3.48) directly.

Problems

Reminder: An asterisk by a problem number indicates that the problem is discussed, or its answer given, in the Answers section at the back of the book.

***3.1** (Section 3.1). Two students are told to measure the rate of emission of α particles from a certain radioactive sample. Student A counts for two minutes, and observes 32 α particles; Student B counts for an hour, and observes 786 α particles. (The sample decays so slowly that the expected rate of emission can be assumed to be constant during the measurements.)

(a) Use Equation (3.2) to calculate the uncertainty in Student A's result, 32, for the number of particles emitted in two minutes.

(b) What is the uncertainty in Student B's result, 786, for the number of particles in a hour?

(c) Each divides his count by the number of minutes to find the *rate*, in decays per minute. What are their answers and their uncertainties? (Although the uncertainty in B's number is larger than that in A's, the uncertainty in B's rate is much smaller than A's; that is, by counting for a longer time, one gets a more precise answer for the rate, just as one would expect.)

3.2 (Section 3.2). A student makes the following measurements:

$$
\begin{aligned}
a &= 5 \pm 1 \text{ cm}; \\
b &= 18 \pm 2 \text{ cm}; \\
c &= 12 \pm 1 \text{ cm}; \\
t &= 3.0 \pm .5 \text{ sec}; \\
m &= 18 \pm 1 \text{ gm}.
\end{aligned}
$$

Using the rules in (3.4) and (3.8), compute the following quantities, with their uncertainties and their percentage uncertainties: $a + b + c$, $a + b - c$, ct, $4a$, $b/2$ (where 4 and 2 have no uncertainty), and mb/t.

***3.3** (Section 3.2). Using the rules in (3.4) and (3.8), compute the following:

(a) $(5 \pm 1) + (8 \pm 2) - (10 \pm 4)$;
(b) $(5 \pm 1) \times (8 \pm 2)$;
(c) $(10 \pm 1)/(20 \pm 2)$;
(d) $2\pi(10 \pm 1)$.

In (d) the numbers 2 and π have no uncertainty.

***3.4** (Section 3.2). With a good stopwatch and some practice, one can measure times ranging from about a second up to many minutes with an uncertainty of 0.1 sec or so. Suppose that we wish to find the period τ of a pendulum with $\tau \approx 0.5$ sec. If we time one oscillation, we will have an uncertainty of about 20 percent; but by timing several successive oscillations, we can do much better, as the following questions illustrate.

(a) If we measure the time for five successive oscillations and get $2.4 \pm .1$ sec, what is our final answer for τ, with its absolute and percent uncertainties? [Remember the rule in (3.9).]
(b) What if we measure 20 oscillations and get $9.4 \pm .1$ sec?
(c) Could the uncertainty in τ be improved indefinitely by timing more and more oscillations?

3.5 (Section 3.2). If t has been found to be $t = 8.0 \pm 0.5$ sec, what are the values and uncertainties of t^2, $1/t$, and $1/t^3$?

***3.6** (Section 3.2). A visitor to a medieval castle decides to measure the depth of a well by dropping a stone and timing its fall. He finds that the time to fall is $t = 3.0 \pm 0.5$ sec. What does he conclude about the depth of the well?

3.7 (Section 3.2). The binomial theorem states that for any number n and any x with $|x| < 1$,

$$(1 + x)^n = 1 + nx + \frac{n(n-1)}{1 \cdot 2}x^2 + \frac{n(n-1)(n-2)}{1 \cdot 2 \cdot 3}x^3 + \cdots.$$

(a) Show that if n is a positive integer, this infinite series terminates (i.e., has only a finite number of terms). Write it down explicitly for the cases $n = 2$ and $n = 3$.
(b) Write down the binomial series for the case $n = -1$. This gives an infinite series for $1/(1 + x)$; when x is small, the first two terms of

this infinite series give a good approximation,

$$1/(1 + x) \approx 1 - x,$$

as quoted in (3.6). Calculate these two expressions for each of the values $x = 0.5$, 0.1, 0.01, and for each calculate the percentage by which the approximation $(1 - x)$ differs from the exact value of $1/(1 + x)$.

***3.8** (Section 3.3). A student measures four lengths:

$$a = 50 \pm 5, \qquad b = 30 \pm 3, \qquad c = 40 \pm 1, \qquad d = 7.8 \pm .3$$

(all in cm), and calculates the three sums $a + b$, $a + c$, $a + d$. Find the resulting uncertainties when the original errors may *not* be independent ("errors add" as in (3.14)), and also when the errors are known to be independent and random ("errors add in quadrature" as in (3.13)). Assuming that the uncertainties are needed with only one significant figure, in which cases can the second uncertainty (that in b, c, or d) be entirely ignored?

3.9 (Section 3.4). Repeat Problem 3.2 assuming that all uncertainties are independent and random, i.e., using addition in quadrature as in the rules (3.16) and (3.18) for error propagation.[9]

***3.10** (Section 3.5). In nuclear physics the energy of a subatomic particle can be measured in various ways. One way is to measure how quickly the particle is stopped by an obstacle such as a piece of lead, and then to use published graphs of energy vs. stopping rate. Figure 3.7 shows such a graph for photons (the particles of light) in lead. The vertical axis shows the photons' energy E in MeV (millions of electron volts); and the horizontal axis shows the corresponding absorption coefficient μ in cm^2/gm. (The precise definition of this coefficient need not concern us here; μ is simply a suitable measure of how quickly the photon is stopped in the lead.) From this graph one can obviously find the energy E of a photon as soon as one knows its absorption coefficient μ.

(a) A student observes a beam of photons (all with the same energy) and finds that their absorption coefficient in lead is $\mu = .10 \pm .01\ \text{cm}^2/\text{gm}$. Using the graph, find what is the energy E of the photons and the uncertainty δE. (You may find it helpful to draw on the graph the lines connecting the various points of interest, as was done in Figure 3.3.)

[9] Addition in quadrature can often be done in the head with sufficient accuracy. When a calculator is used, note that the conversion from rectangular to polar coordinates automatically calculates $\sqrt{x^2 + y^2}$ for any given x and y.

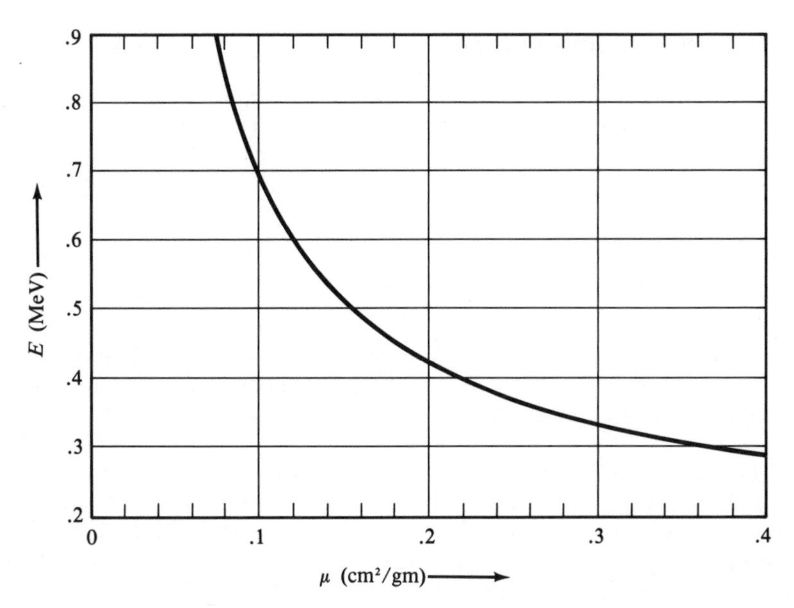

Figure 3.7. Energy E vs. absorption coefficient μ for photons in lead.

(b) What conclusions would the student have drawn if $\mu = .22 \pm .01$ cm^2/gm?

*__3.11__ (Section 3.5).

(a) An angle θ is measured as 125 ± 2 degrees, and this value is used to compute $\sin\theta$. Using the rule in (3.23), calculate $\sin\theta$ and its uncertainty.

(b) If a is measured as $a_{\text{best}} \pm \delta a$, and this value used to compute $f(a) = e^a$, what are f_{best} and δf? If $a = 3.0 \pm 0.1$, what is e^a and its uncertainty?

(c) Repeat the whole of part (b) for the function $f(a) = \ln a$.

3.12 (Section 3.6). Calculate the following quantities in steps as described in Section 3.6. (Assume all errors are independent and random.)

(a) $(12 \pm 1) \times [(25 \pm 3) - (10 \pm 1)]$

(b) $\sqrt{16 \pm 4} + (3.0 \pm 0.1)^3(2.0 \pm 0.1)$

(c) $(20 \pm 2)e^{-(1.0 \pm 0.1)}$

3.13 (Section 3.7). Review the discussion of the simple pendulum in Section 3.7. In a real experiment, one should measure the period T for various different lengths l, and hence obtain various different values of g for comparison. With a little thought, one can organize all data and calculations so that they appear in a single convenient table, as in Table 3.2. Using Table 3.2 (or some other arrangement that you prefer), calculate

Table 3.2. Finding g with a pendulum.

l (cm) all $\pm .1$	T (sec) all $\pm .001$	g (cm/sec^2)	$\delta l/l$ (%)	$\delta T/T$ (%)	$\delta g/g$ (%)	answer: $g \pm \delta g$
93.8	1.944	980	.1	.05	.14	980 ± 1.4
70.3	1.681					
45.7	1.358					
21.2	0.922					

g and its uncertainty δg for the four pairs of data shown. Comment on the variation of δg as l gets smaller. (The answers given for the first pair of data will let you check your method of calculation.)

***3.14** (Section 3.7). Review the measurement of the refractive index of glass in Section 3.7. Using a table similar to Table 3.1, calculate the refractive index n and its fractional uncertainty for the data in Table 3.3. Comment on the variation in the uncertainty. (All angles are measured in degrees; i is the angle of incidence, r that of refraction.)

Table 3.3. Refractive index data (in degrees).

i (all ± 1)	10	20	30	50	70
r (all ± 1)	6	13	19	29	38

3.15 (Section 3.8). Review the experiment of Section 3.8 in which a cart is rolled down an incline of slope θ.

(a) If the cart's wheels are smooth and light, then the expected acceleration is $g \sin \theta$. If θ is measured as $\theta = 5.4 \pm .1$ degrees, what are the expected acceleration and its uncertainty?

Table 3.4. Acceleration experiment.[a]

t_1 (sec) all $\pm .001$	t_2 (sec) all $\pm .001$	$\frac{1}{t_1^2}$	$\frac{1}{t_2^2}$	$\frac{1}{t_2^2} - \frac{1}{t_1^2}$	a (cm/sec^2)
$.054 \pm 2\%$	$.031 \pm 3\%$	343 ± 14	1040 ± 62	698 ± 64	87 ± 8
.038	.027				
.025	.020				

[a] The first line of data is that already used in Section 3.8. The first two columns show the measured times t_1 and t_2. The uncertainty in all these is .001 sec, which can be converted immediately into a percentage uncertainty for each time.

(b) If the experiment is repeated, giving the cart various pushes at the top of the slope, then, as usual, the data and all calculations can be recorded in a single table, as in Table 3.4. Using Equation (3.33) for the acceleration (and the same value $l^2/2s = .125$ cm $\pm$ 2 percent as before), calculate a and δa for the data shown. Are the results consistent with the expected constancy of a, and with the expected value $g \sin \theta$ of part (a)? Would it be worth pushing the cart harder to check the constancy of a at even higher speeds? Explain.

***3.16** (Section 3.9). The partial derivative $\partial q/\partial x$ of $q(x, y)$ is obtained by differentiating q with respect to x while treating y as a constant. Write down the partial derivatives $\partial q/\partial x$ and $\partial q/\partial y$ for the three functions

(a) $q(x, y) = x + y$,

(b) $q(x, y) = xy$,

(c) $q(x, y) = x^2y^3$.

***3.17** (Section 3.9). The crucial approximation used in Section 3.9 relates the value of the function q at the point $(x + u, y + v)$ to that at the nearby point (x, y):

$$q(x + u, y + v) \approx q(x, y) + \frac{\partial q}{\partial x} u + \frac{\partial q}{\partial y} v \tag{3.49}$$

when u and v are small. Verify explicitly that this is a good approximation for the three functions of Problem 3.16. That is, for each of these three functions, write both sides of Equation (3.49) exactly, and show that they are approximately equal when u and v are small. For example, if $q(x, y) = xy$, then the left side of Equation (3.49) is

$$(x + u)(y + v) = xy + uy + xv + uv.$$

As you will show, the right side of (3.49) is

$$xy + yu + xv.$$

If u and v are small, then uv can be neglected in the first expression, and the two expressions are approximately equal.

3.18 (Section 3.9).

(a) For the function $q(x, y) = xy$, write down the partial derivatives $\partial q/\partial x$ and $\partial q/\partial y$. Suppose that we measure x and y with uncertainties δx and δy, and then calculate $q(x, y)$. Using the general rules in (3.47) and (3.48), write down the uncertainty δq both for the case when δx and δy are independent and random, and for the case when they are not. Divide through by $|q| = |xy|$, and show that you recover the simple rules (3.18) and (3.19) for the fractional uncertainty in a product.

(b) Repeat part (a) for the function $q(x, y) = x^n y^m$, where n and m are known, fixed numbers.

(c) What do Equations (3.47) and (3.48) become when $q(x)$ depends on only one variable?

***3.19** (Section 3.9). If we measure three independent quantities x, y, z, and then calculate a function like $q = (x + y)/(x + z)$, then, as was discussed at the beginning of Section 3.9, a step-by-step calculation of the uncertainty in q may overestimate the uncertainty δq.

(a) Consider the measured values $x = 20 \pm 1$, $y = 2$, $z = 0$; and, for simplicity, suppose that δy and δz are negligible. Calculate the uncertainty δq correctly, using the general rule in (3.47), and compare the result with what you would get if you were to calculate δq in steps.

(b) Do the same for the values $x = 20 \pm 1$, $y = -40$, $z = 0$. Explain any differences between parts (a) and (b).

CHAPTER 4

Statistical Analysis of Random Uncertainties

We have seen that one of the best ways to assess the reliability of a measurement is to repeat it several times and to examine the various different values obtained. In this chapter and Chapter 5 we describe statistical methods for analysing measurements in this way.

As has already been mentioned, not all types of experimental uncertainty can be assessed by statistical analysis based on repeated measurements. For this reason uncertainties are classified into two groups: the *random* uncertainties, which *can* be treated statistically; and the *systematic* uncertainties, which *cannot*. This distinction is described in Section 4.1. For most of the remainder of this chapter, we confine our attention to random uncertainties. In Section 4.2 we introduce, without formal justification, two important definitions related to a series of measured values $x_1, \ldots, x_N$, all of some single quantity x. First we define the *average* or *mean* $\bar{x}$ of $x_1, \ldots, x_N$. Under suitable conditions, the mean $\bar{x}$ is the best estimate of x based on the measured values $x_1, \ldots, x_N$. We then define the *standard deviation* of $x_1, \ldots, x_N$. This is denoted σ_x, and characterizes the average uncertainty in the separate measured values $x_1, \ldots, x_N$. In Section 4.3 we give an example of the use of the standard deviation.

In Section 4.4 we introduce the important notion of the *standard deviation of the mean.* This is denoted $\sigma_{\bar{x}}$, and characterizes the uncertainty in the mean $\bar{x}$ as the best estimate for x. In Section 4.5 we give some examples of the standard deviation of the mean. Finally, in Section 4.6 we return to the vexing problem of systematic errors.

Nowhere in this chapter do we attempt a complete justification of the methods described. Our main aim is to introduce the basic formulas and to describe how they are used. In Chapter 5 we will give proper justifications, based on the important idea of the normal distribution curve.

4.1. *Random and Systematic Errors*

Experimental uncertainties that can be revealed by repeating the measurements are called *random* errors; those that cannot be revealed in this way are called *systematic*. To illustrate this distinction, let us consider some examples. Suppose first that we time a revolution of a steadily rotating turntable. One source of error will be our reaction time in starting and stopping the watch. If our reaction time were always exactly the same, these two delays would cancel one another. In practice, however, our reaction time will vary. We may delay more in starting, and so underestimate the time of a revolution; or we may delay more in stopping, and so overestimate the time. Since either possibility is equally likely, the sign of the effect is *random*. If we repeat the measurement several times, we will sometimes overestimate and sometimes underestimate. Thus our variable reaction time will show up as a variation of the answers found. By analysing the spread in results statistically, we can get a very reliable estimate of this kind of error.

On the other hand, if our stopwatch is running consistently slow, then all our times will be underestimates, and no amount of repetition (with the same watch) will reveal this source of error. This kind of error is called *systematic*, because it always pushes our result in the same direction. (If the watch runs slow, we always underestimate; if the watch runs fast, we always overestimate.) Systematic errors cannot be discovered by the kind of statistical analysis that we are contemplating here.

As a second example of random and systematic errors, suppose we have to measure some well-defined length with a ruler. One source of uncertainty will be the need to interpolate between scale markings; and this uncertainty is probably random. (When interpolating, we are probably just as likely to overestimate as to underestimate.) But there is also the possibility that our ruler has become distorted; and this source of uncertainty would probably be systematic. (If the ruler has stretched, we always underestimate; if it has shrunk, we always overestimate.)

Just as in these two examples, almost all measurements are subject to both random and systematic uncertainties. You should have no difficulty in finding more examples. In particular, notice that common sources of random uncertainties are small errors of judgment by the observer (as when interpolating), small disturbances of the apparatus (like mechanical vibrations), problems of definition, and several others. Perhaps the most obvious cause of systematic error is the miscalibration of instruments, like the watch that runs slow, the ruler that has been stretched, or a meter that is improperly zeroed.

The distinction between random and systematic errors is not always clear-cut. For example, if you move your head from side to side in front of a typical meter (such as an ordinary clock), the reading on the meter changes. This effect is called *parallax*, and means that a meter can be read correctly only if you position yourself directly in front of it. Even if you are a careful experimenter, you cannot always position your eye *exactly* in front of the meter; consequently your measurements will have a small uncertainty due to parallax, and this uncertainty will probably be random. On the other hand, a careless experimenter who places a meter to one side of his seat and forgets to worry about parallax will introduce a systematic error into all his readings. Thus the same effect, parallax, can produce random uncertainties in one case, and systematic uncertainties in another.

The treatment of random errors is quite different from that of systematic errors. The statistical methods described in the following sections give a reliable estimate of the random uncertainties, and, as we shall see, provide a well-defined procedure for reducing them. On the other hand, systematic uncertainties are hard to evaluate, and even to detect. The experienced scientist has to learn to anticipate the possible sources of systematic error, and to make sure that all systematic errors are much less than the required precision. Doing so will involve, for example, checking the meters against accepted standards, and correcting them or buying better ones if necessary. Unfortunately, in the beginning physics laboratory, such checks are rarely possible; so the treatment of systematic errors is often awkward. We will discuss this further in Section 4.6. For now, we will discuss experiments in which all sources of systematic error have been identified and made much smaller than the required precision.

4.2. The Mean and Standard Deviation

Suppose we need to measure some quantity x, and have identified all sources of systematic error and reduced them to a negligible level. Since all remaining sources of uncertainty are random, we should be able to detect them by repeating the measurement several times. We might, for example, make the measurement five times and find the results

$$71, 72, 72, 73, 71, \tag{4.1}$$

(where, for convenience, we have omitted any units).

The first question that we address is as follows: Given the five measured values (4.1), what should we take for our best estimate x_{best} of the quantity x? It seems reasonable that our best estimate would be the *average* or *mean* $\bar{x}$ of the five values found, and in Chapter 5 we will prove that this is normally so. Thus

$$\begin{aligned} x_{\text{best}} &= \bar{x} \\ &= \frac{71 + 72 + 72 + 73 + 71}{5} \\ &= 71.8. \end{aligned} \tag{4.2}$$

Here the second line is simply the definition of the mean $\bar{x}$ for the numbers at hand.[1]

More generally, suppose we make N measurements of the quantity x (all using the same equipment and procedures), and find the N values

$$x_1, x_2, \ldots, x_N. \tag{4.3}$$

Once again, the best estimate for x is usually the average of $x_1, \ldots, x_N$. That is,

$$x_{\text{best}} = \bar{x}, \tag{4.4}$$

where

$$\begin{aligned} \bar{x} &= \frac{x_1 + x_2 + \cdots + x_N}{N} \\ &= \frac{\sum x_i}{N}. \end{aligned} \tag{4.5}$$

In the last line we have introduced the useful sigma notation, according to which

$$\sum_{i=1}^{N} x_i = \sum_i x_i = \sum x_i = x_1 + x_2 + \cdots + x_N;$$

the second and third expressions here are common abbreviations, which we will use when there is no danger of confusion.

The concept of the average or mean is almost certainly familiar to most readers. Our next concept, that of the *standard deviation*, is probably less so. The standard deviation of the measurements $x_1, \ldots, x_N$ is an estimate

[1] In this age of pocket calculators, it is perhaps worth pointing out that an average like (4.2) is easily calculated in one's head. Since all the numbers are in the seventies, the same must be true of the average. All that remains is to average the numbers 1, 2, 2, 3, 1 in the units place. These obviously average to $\frac{9}{5} = 1.8$, and our answer is $\bar{x} = 71.8$.

Table 4.1. Calculation of deviations.

Trial number, i	Measured value, x_i	Deviation, $d_i = x_i - \bar{x}$
1	71	-0.8
2	72	0.2
3	72	0.2
4	73	1.2
5	71	-0.8
	$\bar{x} = 71.8$	$\bar{d} = 0.0$

of the *average uncertainty of the measurements* $x_1, \ldots, x_N$, and is arrived at as follows.

Given that the mean $\bar{x}$ is our best estimate of the quantity x, it is natural to consider the difference $x_i - \bar{x} = d_i$. This difference, often called the *deviation* (or residual) of x_i from $\bar{x}$, tells us *how much the* ith *measurement* x_i *differs from the average* $\bar{x}$. If the deviations $d_i = x_i - \bar{x}$ are all very small, then our measurements are all close together and are presumably very precise. If some of the deviations are large, then our measurements are obviously not so precise.

To be sure we understand the idea of the deviation, let us calculate the deviations for the set of five measurements reported in (4.1). These can be listed as shown in Table 4.1. Notice that the deviations are not (of course) all the same size; d_i is small if the ith measurement x_i happens to be close to $\bar{x}$, but d_i is large if x_i is far from $\bar{x}$. Notice also that some of the d_i are positive and some negative, since some of the x_i are bound to be higher than the average $\bar{x}$, and some are bound to be lower.

To estimate the average reliability of the measurements $x_1, \ldots, x_5$, we might naturally try averaging the deviations d_i. Unfortunately, as a glance at Table 4.1 shows, the average of the deviations is zero. In fact, this will be the case for any set of measurements $x_1, \ldots, x_N$, since the definition of the average $\bar{x}$ ensures that $d_i = x_i - \bar{x}$ is sometimes positive and sometimes negative in just such a way that $\bar{d}$ is zero (see Problem 4.3). Obviously, then, the average of the deviations is not a useful way to characterize the reliability of the measurements $x_1, \ldots, x_N$.

The best way to avoid this annoyance is to *square* all the deviations, which will create a set of *positive* numbers, and then average these numbers.[2] If we then take the square root of the result, we obtain a quantity

[2] Another possibility would be to take the absolute values $|d_i|$ and average them; but the average of the d_i^2 proves to be more useful. The average of the $|d_i|$ is sometimes (misleadingly) called the *average deviation*.

with the same units as x itself. This number is called the *standard deviation* of $x_1, \ldots, x_N$, and is denoted σ_x:

$$\sigma_x = \sqrt{\frac{1}{N}\sum_{i=1}^{N}(d_i)^2} = \sqrt{\frac{1}{N}\sum_{i=1}^{N}(x_i - \bar{x})^2}. \tag{4.6}$$

With this definition, the standard deviation can be described as the *root mean square* (or R.M.S.) deviation of the measurements $x_1, \ldots, x_N$. It proves to be a useful way to characterize the reliability of the measurements. (As we will discuss shortly, the definition (4.6) is sometimes modified by replacing the denominator N by $N - 1$.)

To calculate the standard deviation σ_x as defined by (4.6), we must compute the deviations d_i, square them, average these squares, and then take the square root of the result. For the five measurements of Table 4.1, we could calculate σ_x as in Table 4.2.

Table 4.2. Calculation of the standard deviation.

Trial number, i	Measured value, x_i	Deviation, $d_i = x_i - \bar{x}$	d_i^2
1	71	$-.8$	.64
2	72	.2	.04
3	72	.2	.04
4	73	1.2	1.44
5	71	$-.8$	.64
	$\bar{x} = 71.8$		$\sum d_i^2 = 2.80$

Summing the numbers d_i^2 in the fourth column of Table 4.2 and dividing by 5, we obtain the quantity σ_x^2 (often called the *variance* of the measurements),

$$\sigma_x^2 = \frac{1}{N}\sum d_i^2 = \frac{2.80}{5} = .56. \tag{4.7}$$

Taking the square root, we find the standard deviation

$$\sigma_x \approx .7. \tag{4.8}$$

Thus the average uncertainty of the five measurements 71, 72, 72, 73, 71 is about .7.

Unfortunately, there is an alternative definition of the standard deviation. There are theoretical arguments for replacing the factor N in (4.6) by $(N - 1)$ and defining the standard deviation σ_x of $x_1, \ldots, x_N$ as

$$\sigma_x = \sqrt{\frac{1}{N-1}\sum d_i^2} = \sqrt{\frac{1}{N-1}\sum(x_i - \bar{x})^2}. \tag{4.9}$$

We will not try here to prove that definition (4.9) of σ_x is better than (4.6), except to say that the new "improved" definition is obviously a little larger than the old one (4.6), and that this corrects a tendency for (4.6) to understate the uncertainty in the measurements $x_1, \ldots, x_N$, especially if the number of measurements N is small. One can understand this tendency by considering the extreme (and absurd) case that $N = 1$ (i.e., we make only one measurement). Here the average $\bar{x}$ is equal to our one reading x_1, and the one deviation is automatically zero. Therefore the definition (4.6) gives the absurd result $\sigma_x = 0$. On the other hand, the definition (4.9) gives $0/0$; that is, with definition (4.9), σ_x is undefined, which correctly reflects our total ignorance of the uncertainty after just one measurement. The definition (4.6) is sometimes called the *population standard deviation*, and (4.9) the *sample standard deviation.*

The difference between the two definitions (4.6) and (4.9) is almost always numerically insignificant. One should always repeat a measurement many times (at least five, and preferably many more). Even if we make only five measurements ($N = 5$), the difference between $\sqrt{N} = 2.2$ and $\sqrt{N-1} = 2$ is, for most purposes, insignificant. For example, if we recalculate the standard deviation (4.8) using the "improved" definition (4.9), we obtain $\sigma_x = .8$ instead of .7, not a very important difference. Nevertheless, you need to know that both definitions exist. It is probably best always to use the more conservative (i.e., larger) definition (4.9), but, in any event, a laboratory report should state clearly which definition is being used, so that the reader can check the calculations.

4.3. *The Standard Deviation as the Uncertainty in a Single Measurement*

We have said that the standard deviation σ_x characterizes the average uncertainty of the measurements $x_1, \ldots, x_N$ from which it was calculated. In Chapter 5 we will justify this by proving the following more precise statement. If our measurements are normally distributed, and if we were to repeat our measurement of x many more times (always using the same equipment), then about 70 percent of our measurements[3] would lie within a distance σ_x on either side of $\bar{x}$; that is, 70 percent of our measurements would lie in the range $\bar{x} \pm \sigma_x$.

We can rephrase this result as follows. Suppose, as before, that we obtain the values $x_1, \ldots, x_N$ and compute $\bar{x}$ and σ_x. If we then make one more measurement (using the same equipment), there is a 70 percent *probability* that the new measurement will be within σ_x of $\bar{x}$. Now, if the original number of measurements N was large, then $\bar{x}$ should be a very reliable estimate for the actual value of x. Therefore we can say that *there is a 70 percent probability that a single measurement* (using the same equipment) *will be within σ_x of the actual value.* Clearly σ_x means exactly what we have used the term "uncertainty" to mean in the preceding chapters. If we make one measurement of x using this equipment, then the uncertainty associated with this measurement can be taken to be $\delta x = \sigma_x$; and with this choice we are 70 percent confident that our measurement is within δx of the correct answer.

To illustrate the application of these ideas, suppose we are given a box full of similar springs and told to measure their spring constants k. We might measure the spring constants by loading each spring and observing the resulting extension or, perhaps better, by suspending a mass from each spring and timing its oscillations. Whatever method we choose, we will need to know k and its uncertainty δk for each spring, but it would be hopelessly time-consuming to repeat our measurements many times for each spring. Instead we reason as follows. If we measure k for the first spring several, (say, 10 or 20) times, then the mean of these measurements should give a good estimate of k for the first spring. More important here, the standard deviation σ_k of these 10 or 20 measurements provides us with an estimate of the uncertainty in our method for measuring k. Provided our springs are all reasonably similar and we use the same method to measure each one, we can reasonably expect the same uncertainty in each measurement.[4] Thus for each subsequent spring we need to make only one measurement, and we can immediately state that the uncertainty δk is the standard deviation σ_k measured for the first spring, with a 70 percent confidence that our answer is within σ_k of the correct value.

[3] As we will see, the exact number is 68.27 . . . percent, but it is obviously absurd to state this kind of number so precisely.

[4] If some springs are very different from the first, then our uncertainty in measuring them may be different. Thus if the springs differ a lot, we would need to check our uncertainty by making many measurements for each of a few different springs.

To illustrate these ideas numerically, we can imagine making 10 measurements on the first spring, and obtaining the following measured values of k (in newtons/meter):

$$86, 85, 84, 89, 86, 88, 88, 85, 83, 85. \tag{4.10}$$

From these we can immediately calculate $\bar{k} = 85.9$ N/m and, using the definition (4.9),

$$\sigma_k = 1.9 \text{ N/m} \tag{4.11}$$

$$\approx 2 \text{ N/m}. \tag{4.12}$$

The uncertainty in any one measurement of k is therefore about 2 N/m. If we now measure the second spring once and obtain the answer $k =$ 71 N/m, we can without further ado take $\delta k = \sigma_k = 2$ N/m, and state with 70 percent confidence that k lies in the range

$$k \text{ for second spring} = 71 \pm 2 \text{ N/m}. \tag{4.13}$$

4.4. *The Standard Deviation of the Mean*

If $x_1, \ldots, x_N$ are the results of N measurements of the same quantity x, then, as we have seen, our best estimate for the quantity x is their mean, $\bar{x}$. We have also seen that the standard deviation σ_x characterizes the average uncertainty of the separate measurements $x_1, \ldots, x_N$. However, our answer $x_{\text{best}} = \bar{x}$ represents a judicious combination of all N measurements, and there is every reason to think it will be more reliable than any one of the measurements considered separately. In Chapter 5 we will prove that this is so; the uncertainty in the final answer $x_{\text{best}} = \bar{x}$ turns out to be the standard deviation σ_x *divided by* $\sqrt{N}$. This quantity is called the *standard deviation of the mean*, and is denoted $\sigma_{\bar{x}}$:

$$\sigma_{\bar{x}} = \sigma_x/\sqrt{N}. \tag{4.14}$$

(Other common names for this are *standard error* and *standard error of the mean.*) Thus, based on the N measured values $x_1, \ldots, x_N$, we can state our final answer for the value of x as

$$(\text{value of } x) = x_{\text{best}} \pm \delta x$$

where $x_{best} = \bar{x}$, the mean of $x_1, \ldots, x_N$, and δx is the standard deviation of the mean,

$$\delta x = \sigma_{\bar{x}} = \sigma_x/\sqrt{N}. \tag{4.15}$$

As an example, we can consider the ten measurements reported in (4.10). These were ten measurements of the spring constant k of one spring. As we already saw, the mean of these values is $\bar{k} = 85.9$ N/m and the standard deviation is $\sigma_k = 1.9$ N/m. Therefore the standard deviation of the mean is

$$\sigma_{\bar{k}} = \sigma_k/\sqrt{10} = 0.6 \text{ N/m}, \tag{4.16}$$

and our final answer, based on these ten measurements, would be that the spring has

$$k = 85.9 \pm 0.6 \text{ newtons/meter}. \tag{4.17}$$

When you give an answer like this, it is important to state clearly what the numbers are—namely, the mean and the standard deviation of the mean—so that the reader will be able to judge their significance.

An important feature of the standard deviation of the mean, $\sigma_{\bar{x}} = \sigma_x/\sqrt{N}$, is the factor $\sqrt{N}$ in the denominator. The standard deviation σ_x represents the average uncertainty in the individual measurements $x_1, \ldots, x_N$. Thus if we were to make some more measurements (using the same technique) the standard deviation σ_x would not change appreciably. On the other hand, the standard deviation of the mean, $\sigma_x/\sqrt{N}$, would slowly decrease as we increase N. This is just what we would expect. If we make more measurements before computing an average, we would naturally expect the final result to be more reliable, and this is just what the denominator $\sqrt{N}$ in (4.15) guarantees. This provides one obvious way to improve the precision of our measurements.

Unfortunately, the factor $\sqrt{N}$ grows rather slowly as we increase N. For example, if we wish to improve our precision by a factor of 10 simply by increasing the number of measurements N, we will have to increase N by a factor of 100—a daunting prospect, to say the least! Furthermore, we are for the moment neglecting systematic errors, and these are *not* reduced by increasing the number of measurements. Thus, in practice, if you want to increase your precision appreciably, you will probably do better to improve your technique than to rely merely on increased numbers of measurements.

4.5. Examples

As a first, simple application of the standard deviation of the mean, imagine that we have to measure very accurately the area A of a rectangular plate about 2.5 cm × 5 cm. We first find the best available measuring device, which might be a vernier caliper, and then make several measurements of the length l and breadth b of the plate. To allow for irregularities in the sides, we make our measurements at several different positions, and to allow for small defects in the instrument, we use several different calipers (if available). We might make ten measurements each of l and b, and obtain the results shown in Table 4.3.

Table 4.3. Length and breadth (in mm).

	Measured values	Mean	SD	SDOM
l	24.25, 24.26, 24.22, 24.28, 24.24 24.25, 24.22, 24.26, 24.23, 24.24	$\bar{l} = 24.245$	$\sigma_l = .019$	$\sigma_{\bar{l}} = .006$
b	50.36, 50.35, 50.41, 50.37, 50.36 50.32, 50.39, 50.38, 50.36, 50.38	$\bar{b} = 50.368$	$\sigma_b = .024$	$\sigma_{\bar{b}} = .008$

Using the ten observed values of l, we can quickly calculate the mean $\bar{l}$, the standard deviation σ_l, and the standard deviation of the mean $\sigma_{\bar{l}}$, as shown in the columns labeled mean, SD, and SDOM. In the same way we can calculate $\bar{b}$, σ_b, and $\sigma_{\bar{b}}$. Before doing any further calculations, you should examine these results to see if they seem reasonable. For example, the two standard deviations σ_l and σ_b are supposed to be the average uncertainty in the measurements of l and b. Since l and b were measured in exactly the same way, it would be rather surprising if σ_l and σ_b differed significantly from each other, or from what we judge to be a reasonable uncertainty for the measurements.

Having convinced ourselves that the results so far are reasonable, we can quickly finish our calculations. Our best estimate for the length is the mean $\bar{l}$ and our uncertainty is the SDOM, $\sigma_{\bar{l}}$; so our final value for l is

$$l = 24.245 \pm .006 \text{ mm (or } .025\%\text{)},$$

the number in parenthesis being the percentage uncertainty. Similarly, our value for b is

$$b = 50.368 \pm .008 \text{ mm (or } .016\%\text{)}.$$

Finally, our best estimate for the area $A = lb$ is the product of these values, with a fractional uncertainty given by the quadratic sum of those in l and b (assuming that the errors are independent):

$$\begin{aligned} A &= (24.245 \text{ mm} \pm .025\%) \times (50.368 \text{ mm} \pm .016\%) \\ &= 1221.17 \text{ mm}^2 \pm .03\% \\ &= 1221.2 \pm .4 \text{ mm}^2. \end{aligned} \tag{4.18}$$

To arrive at the answer (4.18) for A, we calculated the averages $\bar{l}$ and $\bar{b}$, each with an uncertainty equal to the standard deviation of its mean. We then calculated the area A as the product of $\bar{l}$ and $\bar{b}$, and found the uncertainty by propagation of errors. We could have proceeded differently. For instance, we could have multiplied the first measured value of l by the first value of b to give a first answer for A. Continuing in this way we could have calculated ten answers for A, and then have subjected these 10 answers to statistical analysis, calculating $\bar{A}$, σ_A, and finally $\sigma_{\bar{A}}$. However, if the errors in l and b are independent and random, and if we make enough measurements, this alternative procedure will (it can be shown) produce the same result as the first one.[5]

As a second example, we consider a case where one cannot conveniently apply a statistical analysis to the direct measurements, but can to the final answers. Suppose we wish to measure the spring constant k of a spring by timing the oscillations of a mass m fixed to its end. It is well-known from elementary mechanics that the period for such oscillations is $T = 2\pi\sqrt{m/k}$. Thus by measuring T and m we can find k as

$$k = 4\pi^2 m/T^2. \tag{4.19}$$

The simplest way to find k is to take a single accurately known mass m, and make several careful measurements of T. However, for various reasons it may be more interesting to time T for various *different* masses m. (For example, in this way we could check that $T \propto \sqrt{m}$, as well as measure k.) We might then get a set of readings like those on the first two lines of Table 4.4.

It obviously makes no sense to average the various different masses in the top line (nor the times in the second line), since they are *not* different measurements of the same quantity. Nor can we learn anything about the

[5] There is a certain illogic to the second procedure, since there is no particular reason to associate the first measurement of l with the first measurement of b. Indeed, we might have measured l eight times, and b 12 times; then we couldn't pair off values. Thus our first procedure is logically preferable.

Table 4.4. Measurement of spring constant k.

mass m (kg)	.513	.581	.634	.691	.752	.834	.901	.950
period T (sec)	1.24	1.33	1.36	1.44	1.50	1.59	1.65	1.69
$k = 4\pi^2 m/T^2$	13.17	12.97	etc.					

uncertainty in our measurements by comparing the different values of m. On the other hand, we can combine each value of m with its corresponding period T and calculate k, as in the final line of Table 4.4. Our answers for k in the bottom line *are* all measurements of the same quantity, and so can be subjected to statistical analysis. In particular, our best estimate for k is the mean, $\bar{k} = 13.16\ N/m$, and our uncertainty is the standard deviation of the mean, $\sigma_{\bar{k}} = .06\ N/m$ (see Problem 4.12). Thus the final answer, based on the data of Table 4.4, is

$$\text{spring constant } k = 13.16 \pm .06 \text{ N/m}. \tag{4.20}$$

If we had formed reasonable estimates of the uncertainties in our original measurements of m and T, we could also have estimated the uncertainty in k by using error propagation, starting from these estimates for δm and δT. In this case, it would be a good idea to compare the final uncertainties in k obtained by the two methods.

4.6. Systematic Errors

In the last few sections, we have been taking for granted that all systematic errors were reduced to a negligible level before serious measurements began. Here we take up again the disagreeable possibility that there are appreciable systematic errors. In the example just discussed, we may have been measuring m with a balance that read consistently high or low, or our timer may have been running consistently fast or slow. Neither of these systematic errors will show up in the comparison of our various answers for the spring constant k. As a result, the standard deviation of the mean, $\sigma_{\bar{k}}$, can be regarded as the *random component* δk_{ran} of the uncertainty δk, but is certainly not the total uncertainty δk. Our problem is to decide how to estimate the *systematic component* δk_{sys}, and then how to combine δk_{ran} and δk_{sys} to give the complete uncertainty δk.

There is no simple theory to tell us what to do about systematic errors. In fact, the only theory of systematic errors is that they must be identified and reduced until they are much less than the required precision. However, in a teaching laboratory this is often not possible. It is often not possible to check a meter against a better one in order to correct it, and still less to buy a new meter to replace an inadequate one. For this reason some teaching laboratories establish a rule that, in the absence of more specific information, meters should be considered to have some definite systematic uncertainty. For example, it might be decided that all stopwatches have up to 0.5 percent systematic uncertainty, all balances up to 1 percent, all voltmeters and ammeters up to 3 percent, and so on.

Given rules of this kind, there are various possible ways to proceed. None of these can really be rigorously justified, and we describe just one approach here. In the last example in Section 4.5, the spring constant $k = 4\pi^2 m/T^2$ was found by measuring a series of values of m and the corresponding values of T. As has been mentioned, a statistical analysis of the various answers for k gives the random component of δk as

$$\delta k_{\text{ran}} = \sigma_{\bar{k}} = .06 \text{ N/m}. \tag{4.21}$$

Suppose now we have been told that the balance used to measure m and the clock used for T have systematic uncertainties up to 1 percent and 0.5 percent, respectively. We can then find the systematic component of δk by propagation of errors, the only question being whether to combine the errors in quadrature or directly. Since the errors in m and T are surely independent, and some cancellation is therefore possible, it is probably reasonable to use the quadratic sum to give[6]

$$\frac{\delta k_{\text{sys}}}{k} = \sqrt{\left(\frac{\delta m_{\text{sys}}}{m}\right)^2 + \left(2\frac{\delta T_{\text{sys}}}{T}\right)^2} \tag{4.22}$$

$$= \sqrt{1 + 1}\% = 1.4\% \tag{4.23}$$

and hence

$$\delta k_{\text{sys}} = (13.16 \text{ N/m}) \times .014$$

$$= .18 \text{ N/m}. \tag{4.24}$$

[6] Whether we should use the quadratic or ordinary sum really depends on what is meant by the statement that the balance has "up to 1 percent systematic uncertainty." If it means the error is *certainly* no more than 1 percent (and likewise for the clock), then direct addition is appropriate, and δk_{sys} is then *certainly* no more than 2 percent. On the other hand, it could be that an analysis of all balances in the laboratory has shown that they follow a normal distribution, with 70 percent of them better than 1 percent reliable (and likewise for the clocks). In this case we can use addition in quadrature as in (4.22) with the usual significance of 70 percent confidence.

Since we now have estimates for both the random and the systematic components of δk, our only remaining problem is to combine them to give δk itself. One can argue that they should be combined in quadrature, to give the total uncertainty

$$\delta k = \sqrt{(\delta k_{\text{ran}})^2 + (\delta k_{\text{sys}})^2} \tag{4.25}$$

$$= \sqrt{(.06)^2 + (.18)^2} \approx .2 \text{ N/m}. \tag{4.26}$$

In this example the systematic uncertainties completely dominate the random ones.

The expression (4.25) for δk cannot really be rigorously justified. Nor is the significance of the answer clear; for example, we probably cannot assert that we have 70 percent confidence that the true answer lies in the range $\bar{k} \pm \delta k$. Nonetheless, the expression does at least provide a reasonable estimate of our total uncertainty, given that our apparatus has systematic uncertainties that we were unable to eliminate. In particular, there is one important respect in which the answer (4.25) is realistic and instructive. We saw in Section 4.4 that the standard deviation of the mean $\sigma_{\bar{k}}$ approaches zero as the number of measurements N is increased. This result suggested that, if you have the patience to make an enormous number of measurements, then you can reduce the uncertainties indefinitely, without having to improve your equipment or technique. We can now see that this is not really so. Increasing N can reduce the *random* component $\delta k_{\text{ran}} = \sigma_{\bar{k}}$ indefinitely. But any given apparatus has *some* systematic uncertainty, which is *not* reduced as we increase N. It is clear from (4.25) that little is gained from further reduction of δk_{ran}, once δk_{ran} is smaller than δk_{sys}. In particular, the total δk can never be made less than δk_{sys}. This simply confirms what we already guessed, that in practice a large reduction of the uncertainty requires improvements in techniques or equipment in order to reduce both the random and the systematic errors in each single measurement.

Problems

Reminder: An asterisk indicates that the problem is discussed, or its answer given, in the Answers section at the back of the book.

***4.1** (Section 4.2). A student measures a quantity x five times, with the results

$$5, 7, 9, 7, 8.$$

Calculate the mean $\bar{x}$ and standard deviation σ_x. (Do the calculation yourself; don't just press the appropriate buttons on your calculator. State which definition of σ_x you use.)

4.2 (Section 4.2). Calculate the mean and standard deviation of the ten measurements reported in (4.10). (The answers are given in the text; but what is important is that you should actually *do* the calculation yourself. You need to decide on some tidy format for your calculations; one possibility is shown in Table 4.2.)

***4.3** (Section 4.2). The mean $\bar{x}$ of N quantities $x_1, \ldots, x_N$ is defined as their sum divided by N; that is, $\bar{x} = (\sum x_i)/N$. The deviation of x_i is the difference $d_i = x_i - \bar{x}$. Show clearly that the mean of the deviations $d_1, \ldots, d_N$ is automatically zero.

If you are not used to the $\sum$ notation, it might help to do this problem both without and with the notation. For example, write out the sum $\sum(x_i - \bar{x})$ as $(x_1 - \bar{x}) + (x_2 - \bar{x}) + \cdots + (x_N - \bar{x})$, and regroup.

***4.4** (Section 4.2). To compute the standard deviation σ_x of N measurements $x_1, \ldots, x_N$, one needs the sum $\sum(x_i - \bar{x})^2$. Prove that this sum can be rewritten as

$$\sum[(x_i - \bar{x})^2] = [\sum(x_i)^2] - N\bar{x}^2. \tag{4.27}$$

(This is a good exercise in using the $\sum$ notation. The result is very useful in practice, and is what all hand calculators use to compute σ_x.)

4.5 (Section 4.2). Recalculate the standard deviation in Problem 4.1 using the identity (4.27).

4.6 (Section 4.3). A student times a pendulum three times and gets the answers 1.6, 1.8, 1.7 (all in sec). What are the mean and standard deviation? (Use the definition (4.9) of the standard deviation.) If the student makes a fourth measurement, what is the probability that this new measurement will lie outside the range 1.6 to 1.8 sec? (Obviously these numbers are chosen to "come out right." In Chapter 5 we will see how to do this kind of problem even when the numbers don't come out right.)

***4.7** (Section 4.3).

(a) Calculate the mean $\bar{t}$ and standard deviation σ_t of the following 30 measurements of a time t (all in sec). You will need a calculator, but you will save a lot of button pushing if you recognize that only the last two digits need averaging, and if you shift the decimal point two places to the right before calculating. If your calculator does not compute standard deviations automatically, you should probably use the identity (4.27).

8.16, 8.14, 8.12, 8.16, 8.18, 8.10, 8.18, 8.18, 8.18, 8.24,
8.16, 8.14, 8.17, 8.18, 8.21, 8.12, 8.12, 8.17, 8.06, 8.10,
8.12, 8.10, 8.14, 8.09, 8.16, 8.16, 8.21, 8.14, 8.16, 8.13.

(b) We have seen that after many measurements we can expect about 70 percent of all values to be within σ_t of $\bar{t}$ (i.e., inside the range $\bar{t} \pm \sigma_t$). In Chapter 5 we will show that we can also expect about 95 percent of all values to be within $2\sigma_t$ of $\bar{t}$ (i.e., inside the range $\bar{t} \pm 2\sigma_t$). For the measurements of part (a), about how many would you expect to lie *outside* the range $\bar{t} \pm \sigma_t$? How many do? Answer the same questions for the number outside $\bar{t} \pm 2\sigma_t$.

4.8 (Section 4.4). Compute the standard deviation of the mean for the five measurements of Problem 4.1. What should be the student's final answer, with its uncertainty, for x?

***4.9** (Section 4.4). Based on the thirty measurements in Problem 4.7, what would be your best estimate for the time involved and its uncertainty, assuming that all uncertainties are random?

4.10 (Section 4.4). After measuring the speed of sound u several times, a student concludes that the standard deviation σ_u of the measurements is $\sigma_u = 10$ m/sec. Assuming that the errors are all random, the student can get any desired precision by making enough measurements and averaging. How many measurements are needed to give a final uncertainty of ± 3 m/sec? How many for an uncertainty of only 0.5 m/sec?

***4.11** (Section 4.5). In Table 4.3 are recorded ten measurements each of the length l and breadth b of a rectangle used to calculate the area $A = lb$. If the measurements were made in pairs (one of l and one of b), then it would be natural to multiply each pair together to give a value of A: the first l times the first b to give a first value of A; and so on. Calculate the resulting ten values of A; the mean, $\bar{A}$; the SD, σ_A; and the SDOM, $\sigma_{\bar{A}}$. Compare your answers for $\bar{A}$ and $\sigma_{\bar{A}}$ with the answer (4.18) obtained by calculating the averages $\bar{l}$ and $\bar{b}$, and then taking A to be $\bar{l}\bar{b}$, with an uncertainty given by error propagation. (For a large number of measurements, the two methods should agree.)

4.12 (Section 4.5). Complete the calculations of the spring constant k in Table 4.4. Then compute $\bar{k}$ and its uncertainty (i.e., the SDOM, $\sigma_{\bar{k}}$).

***4.13** (Section 4.6).

(a) A student measures the speed of sound as $u = f\lambda$, where f is the frequency shown on the dial of an audio oscillator, and λ is the wavelength measured by locating several maxima in a resonant air column. Since there are several measurements of λ, they can be analyzed statistically; and the student concludes that $\lambda = 11.2 \pm .5$ cm. There is only one measurement of $f = 3{,}000$ Hz (the setting on the oscillator) and the student has no way to judge its reliability. The instructor says that the oscillator is "certainly 1 percent reliable"; therefore the student allows for a 1 percent systematic error in f (but none in λ). What is the student's answer for u and its uncertainty? Is the possible 1 percent systematic error from the oscillator calibration important?

(b) If the student's measurement had been $\lambda = 11.2 \pm .1$ cm, and the oscillator calibration were 3 percent reliable, what would the answer have been? Is the systematic error important?

CHAPTER 5

The Normal Distribution

In this chapter we continue our discussion of the statistical analysis of repeated measurements. In Chapter 4 we introduced the important ideas of the mean, the standard deviation, and the standard deviation of the mean; we saw their significance and some of their uses. In this chapter we supply the theoretical justification for these statistical ideas, and we give proofs of several results that were stated without proof in earlier chapters.

The first problem in discussing measurements that are repeated many times is to find a way to handle and display the many values obtained. One convenient method is to use a *distribution* or *histogram*, as described in Section 5.1. In Section 5.2 we introduce the notion of the *limiting distribution*, the distribution of results that would be obtained if the number of measurements became infinitely large. In Section 5.3 we define the *normal distribution* or *Gauss distribution*, which describes the limiting distribution of results for any measurement subject to many small random errors.

Once the mathematical properties of the normal distribution are understood, we can proceed to prove several important results quite easily. In Section 5.4 we prove that, as anticipated in Chapter 4, about 70 percent of all measurements (all of one quantity and all using the same technique) should lie within one standard deviation of the true value. In Section 5.5 we prove the result, used as far back as Chapter 1, that if we make N measurements $x_1, x_2, \ldots, x_N$ of some quantity x, then our best estimate x_{best} based on these values is the mean $\bar{x} = \sum x_i/N$. In Section 5.6 we justify the use of addition in quadrature when propagating errors that are independent and random. In Section 5.7 we prove that the uncertainty of the mean $\bar{x}$, when used as the best estimate of x, is given by the standard deviation of the mean $\sigma_{\bar{x}} = \sigma_x/\sqrt{N}$, as stated in Chapter 4. Finally, in Section 5.8 we discuss how to assign a numerical confidence to experimental results.

The mathematics used in this chapter is a little more advanced than that used up to now. However, the reader who follows carefully the discussion of the normal distribution in Section 5.3 (going through calculations with pencil and paper if necessary) should be able to follow most of the arguments without too much difficulty.

5.1. Histograms and Distributions

It should be clear that the serious statistical analysis of an experiment requires us to make many measurements. Thus our first problem is to devise methods for recording and displaying large numbers of measured values. Suppose, for instance, that we were to make ten measurements of some length x. For example, x might be the distance from a lens to an image formed by the lens. We might obtain the values (all in cm)

$$26, 24, 26, 28, 23, 24, 25, 24, 26, 25. \qquad (5.1)$$

Written in this way, these ten numbers convey fairly little information; and if we were to record many more measurements in this way, the result would be a confusing jungle of numbers. Obviously a better system is called for.

As a first step we can reorganize the numbers (5.1) in ascending order,

$$23, 24, 24, 24, 25, 25, 26, 26, 26, 28. \qquad (5.2)$$

Next, rather than recording the three readings 24, 24, 24, we can simply record that we obtained the value 24 three times. In other words, we can record the *different* values of x obtained, together with the *number* of times each value was found, as in Table 5.1.

Table 5.1.

different values, x_k	23	24	25	26	27	28
number of times found, n_k	1	3	2	3	0	1

Here we have introduced the notation x_k ($k = 1, 2, \ldots$) to denote the various different values found: $x_1 = 23$, $x_2 = 24$, $x_3 = 25$, and so on. And n_k ($k = 1, 2, \ldots$) denotes the number of times the corresponding value x_k was found: $n_1 = 1$, $n_2 = 3$, and so on.

If we record measurements as in Table 5.1, then we can rewrite the definition of the mean $\bar{x}$ in what proves to be a more convenient way. From our old definition we know that

$$\bar{x} = \frac{\sum_i x_i}{N}$$

$$= \frac{23 + 24 + 24 + 24 + 25 + \cdots + 28}{10}. \tag{5.3}$$

This is the same thing as

$$\bar{x} = \frac{23 + (24 \times 3) + (25 \times 2) + \cdots + 28}{10}$$

or, in general

$$\bar{x} = \frac{\sum_k x_k n_k}{N}. \tag{5.4}$$

In the original form (5.3), we sum over *all* the measurements made; in (5.4) we sum over all *different* values obtained, multiplying each value by the number of times it occurred. It is obvious that these two sums are the same, but the form (5.4) proves more useful when we make large numbers of measurements. A sum like that in (5.4) is sometimes called a *weighted sum*, each value x_k being *weighted* by the number of times it occurred, n_k. For later reference, we note that if we add up all the numbers n_k, then we obtain the total number of measurements made, N. That is,

$$\sum_k n_k = N. \tag{5.5}$$

(For example, for Table 5.1 this equation asserts that the sum of the numbers in the bottom line is 10.)

The ideas of the last two paragraphs can be rephrased in a way that is often more convenient. Instead of saying that the result $x = 24$ was obtained three times, we can say that $x = 24$ was obtained in $\frac{3}{10}$ of all our measurements. In other words, instead of using n_k, the *number* of times the result x_k occurred, we introduce the fraction

$$F_k = \frac{n_k}{N}, \tag{5.6}$$

which is the *fraction* of our N measurements that gave the result x_k. The fractions F_k are said to specify the *distribution* of our results, since they describe how our measurements were *distributed* among the different possible values.

In terms of the fractions F_k, we can rewrite the formula (5.4) for the mean $\bar{x}$ in the compact form

$$\bar{x} = \sum_k x_k F_k. \tag{5.7}$$

That is, the mean $\bar{x}$ is just the weighted sum of all the different values x_k obtained, with each x_k weighted by the fraction of times it occurred, F_k.

The result (5.5) implies that

$$\sum_k F_k = 1. \tag{5.8}$$

That is, if we add up the fractions F_k for all possible results x_k, we must get 1. Any set of numbers whose sum is 1 is said to be *normalized*, and the relation (5.8) is therefore called the *normalization condition.*

The distribution of our measurements can be graphically displayed in a *histogram*, as in Figure 5.1. This is just a plot of F_k against x_k, with the different measured values x_k plotted along the horizontal axis, and the fraction of times each x_k was obtained indicated by the height of the vertical bar drawn above x_k. (One can also plot n_k against x_k, but for our purposes the plot of F_k against x_k is more convenient.) Data displayed in

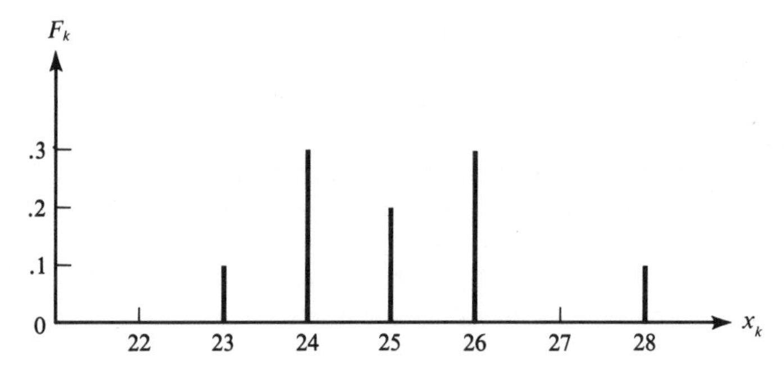

Figure 5.1. Histogram for ten measurements of a length x. The vertical axis shows the fraction of times F_k that each value x_k was observed.

histograms like this can be quickly and easily comprehended, as many writers in newspapers and magazines are aware.

A histogram like that in Figure 5.1 can be called a *bar histogram*, since the distribution of results is indicated by the heights of the vertical bars above the x_k. This kind of histogram is appropriate whenever the values x_k are tidily spaced, with integer values. (For example, students' scores on an exam are usually integers, and are conveniently displayed with a bar histogram.) However, most measurements do not provide tidy integer results, since most physical quantities have a continuous range of possible values. For example, rather than the ten lengths reported in Equation (5.1), you are much more likely to obtain ten values like

$$26.4, 23.9, 25.1, 24.6, 22.7, 23.8, 25.1, 23.9, 25.3, 25.4. \tag{5.9}$$

A bar histogram of these ten values would consist of ten separate bars, all the same height, and would convey comparatively little information. Given measurements like those in (5.9), the best course is to divide the range of values into a convenient number of *intervals* or "*bins*," and to count how many values fall into each "bin." For example, we could count the number of the measurements (5.9) between $x = 22$ and 23, between $x = 23$ and 24, and so on. The results of counting in this way are shown in Table 5.2. (If a measurement happens to fall exactly on the boundary between two bins, then one must decide where to place it. A simple and reasonable course is to assign half a measurement to each of the two bins.)

Table 5.2.

Bin	22 to 23	23 to 24	24 to 25	25 to 26	26 to 27	27 to 28
Observations in bin	1	3	1	4	1	0

The results in Table 5.2 can be plotted in what we can call a *bin histogram*, as shown in Figure 5.2. In this plot the fraction of measurements that fall in each bin is indicated by the area of the rectangle drawn above the bin. Thus the shaded rectangle above the interval from $x = 23$ to $x = 24$ has area $.3 \times 1 = .3$, indicating that $\frac{3}{10}$ of all the measurements fell in this interval. In general we denote the width of the kth bin by Δ_k. (These widths are usually all the same, though they certainly don't have to be.) The height f_k of the rectangle drawn above this bin is chosen so that the area $f_k\Delta_k$ is

$$f_k\Delta_k = \text{fraction of measurements in } k\text{th bin.}$$

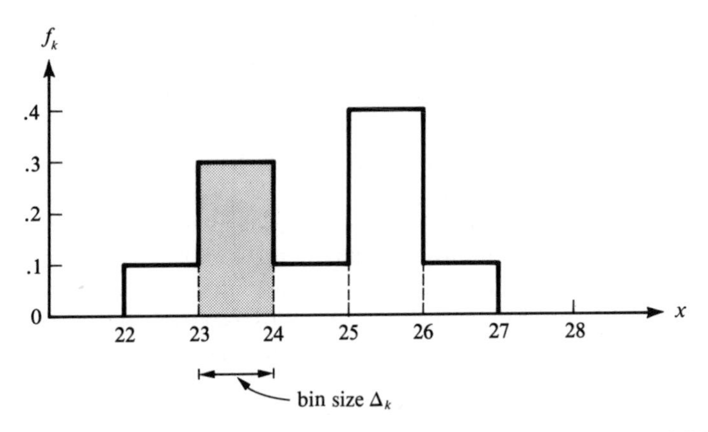

Figure 5.2. Bin histogram showing the fraction of measurements of x that fall in the "bins" 22 to 23, 23 to 24, and so on. The area of the rectangle above each interval gives the fraction of measurements that fall in that interval. Thus the area of the shaded rectangle is .3, indicating that 3/10 of all measurements lie between 23 and 24.

In other words, in a bin histogram the area $f_k\Delta_k$ of the kth rectangle has the same significance as the height F_k of the kth bar in a bar histogram.

Some care is needed in choosing the width Δ_k of the bins for a histogram. If the bins are made much too wide, then all the readings (or almost all) will fall in one bin, and the histogram will be an uninteresting single rectangle. If the bins are made too narrow, then few of them will contain more than one reading, and the histogram will consist of a large number of narrow rectangles almost all of the same height. Clearly, the bin width must be chosen so that there are several readings in each of several bins. Thus when the total number of measurements N is small, we have to choose our bins relatively wide; but if we increase N, then it is usually possible to choose narrower bins.

5.2. *Limiting Distributions*

In most experiments, if one increases the number of measurements, then the histogram begins to take on some definite simple shape. This is clearly visible in Figures 5.3 and 5.4, which show 100 and 1,000 measurements of the same quantity as in Figure 5.2. After 100 measurements, the histogram has become a single peak, which is approximately symmetrical. With 1,000 measurements we have been able to halve the bin size, and the histogram has become quite smooth and regular. These three pictures

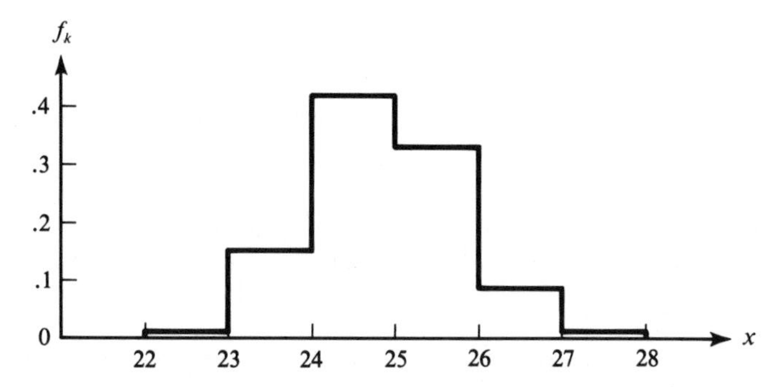

Figure 5.3. Histogram for 100 measurements of the same quantity as in Figure 5.2.

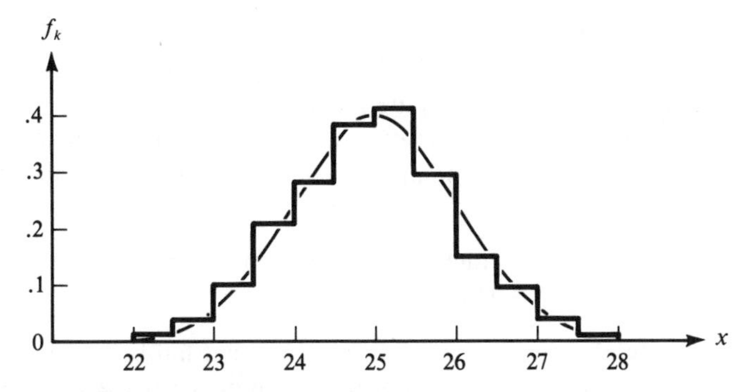

Figure 5.4. Histogram for 1,000 measurements of the same quantity as in Figure 5.3. The broken curve is the limiting distribution.

illustrate an important property of most measurements. As the number of measurements approaches infinity, their distribution approaches some definite, continuous curve. When this happens, the continuous curve is called the *limiting distribution*.[1] Thus for the measurements of Figures 5.2 to 5.4 the limiting distribution appears to be something like the symmetrical bell-shaped curve that has been superimposed on Figure 5.4.

It should be emphasized that the limiting distribution is a theoretical construct, which can never itself be measured exactly. The more measurements we make, the closer our histogram approaches the limiting distribution. But only if we were to make infinitely many measurements and use infinitesimally narrow bins, would we actually obtain the limiting distribution itself. Nevertheless, there is overwhelming evidence that for

[1] Some of the commoner synonyms (or approximate synonyms) for the limiting distribution are: parent distribution, infinite parent distribution, universe distribution, and parent population.

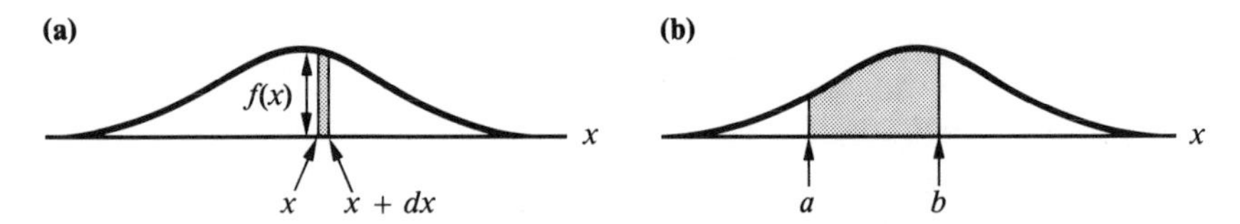

Figure 5.5. A limiting distribution $f(x)$. (a) After very many measurements, the fraction that falls between x and $x + dx$ is the area $f(x)\,dx$ of the narrow strip. (b) The fraction that falls between $x = a$ and $x = b$ is the shaded area.

almost all measurements there *is* a limiting distribution, which our histogram approaches more and more closely as we make more and more measurements.

A limiting distribution, such as the smooth curve in Figure 5.4, defines a function, which we call $f(x)$. The significance of this function is shown by Figure 5.5. As we make more and more measurements of the quantity x, our histogram will eventually be indistinguishable from the limiting curve $f(x)$. Therefore the fraction of measurements that fall in any small interval x to $x + dx$ equals the area $f(x)\,dx$ of the shaded strip in Figure 5.5(a):

$$f(x)\,dx = \text{fraction of measurements that fall between } x \text{ and } x + dx. \tag{5.10}$$

More generally, the number of measurements that fall between any two values a and b is the total area under the graph between $x = a$ and $x = b$ (Figure 5.5b). This area is just the *definite integral* of $f(x)$. Thus we have the important result that

$$\int_a^b f(x)\,dx = \text{fraction of measurements that fall between } x = a \text{ and } x = b. \tag{5.11}$$

It is important to understand the meaning of the two statements (5.10) and (5.11). Both tell us the fraction of measurements expected to lie in some interval, after we make a *very large number of measurements*. Another, and very useful, way to say this is that $f(x)\,dx$ is the *probability* that a single measurement of x will give an answer between x and $x + dx$,

$$f(x)\,dx = \text{probability that any one measurement will will give an answer between } x \text{ and } x + dx. \tag{5.12}$$

Similarly, the integral $\int_a^b f(x)\,dx$ tells us the probability that any one measurement will fall between $x = a$ and $x = b$. We have arrived at the following important conclusion: If we knew the limiting distribution $f(x)$ for the measurement of a given quantity x with a given apparatus, then we would know the probability of obtaining an answer in any interval $a \leqslant x \leqslant b$.

Since the total probability of obtaining an answer anywhere between $-\infty$ and $+\infty$ must be one, the limiting distribution $f(x)$ must satisfy

$$\int_{-\infty}^{\infty} f(x)\,dx = 1. \tag{5.13}$$

This identity is the natural analog of the normalization sum (5.8), $\sum_k F_k = 1$, and a function $f(x)$ satisfying (5.13) is said to be *normalized.*

The reader may be puzzled by the limits $\pm\infty$ in the integral (5.13). These do not mean that we really expect to obtain answers ranging all the way from $-\infty$ to ∞. Quite the contrary. In any real experiment the measurements all fall in some fairly small finite interval. For example, the measurements of Figure 5.4 all lie between $x = 21$ and $x = 29$. Even after infinitely many measurements, the fraction lying outside $x = 21$ to $x = 29$ would be entirely negligible. In other words, $f(x)$ is essentially zero outside this range, and it makes no difference whether the integral (5.13) runs from $-\infty$ to $+\infty$ or 21 to 29. Since we generally don't know what these finite limits are, it is more convenient to leave them as $\pm\infty$.

If the measurement under consideration is very precise, then all the values obtained will be very close to the actual value of x; so the histogram of results, and hence the limiting distribution, will be narrowly peaked, like the solid curve in Figure 5.6. If the measurement is of low precision,

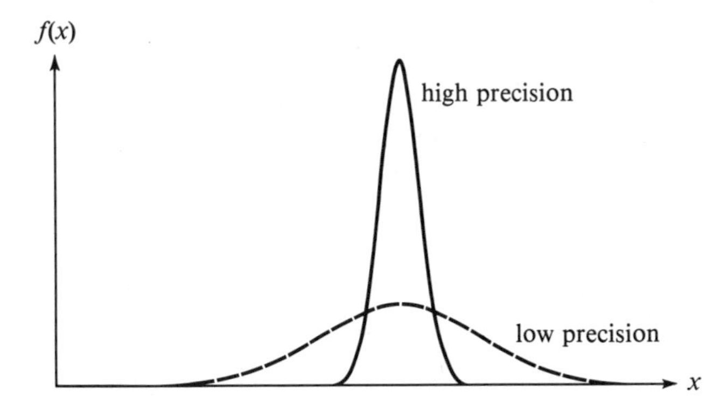

Figure 5.6. Two limiting distributions, one for a high precision measurement, the other for a low precision measurement.

then the values found will be widely spread out, and the distribution will be broad and low, like the dashed curve in Figure 5.6.

The limiting distribution $f(x)$, for measurement of a given quantity x, using a given apparatus, describes how results would be distributed after many, many measurements. Thus if we knew $f(x)$, we could calculate the mean value $\bar{x}$ that would be found after many measurements. We saw in (5.7) that the mean of any number of measurements is the sum of all different values x_k, each weighted by the fraction of times it is obtained,

$$\bar{x} = \sum_k x_k F_k. \tag{5.14}$$

In the present case we have an enormous number of measurements with distribution $f(x)$. If we divide the whole range of values into small intervals x_k to $x_k + dx_k$, then the fraction of values in each interval is $F_k = f(x_k)\,dx_k$ and in the limit that all intervals go to zero, (5.14) becomes

$$\bar{x} = \int_{-\infty}^{\infty} x f(x)\,dx. \tag{5.15}$$

Remember that this formula gives the mean $\bar{x}$ expected after infinitely many trials.

Similarly, we can calculate the standard deviation σ_x obtained after many measurements. Since we are concerned with the limit $N \to \infty$, it makes no difference which definition of σ_x we use, the original (4.6) or the "improved" (4.9) with N replaced by $N - 1$. In either case, when $N \to \infty$, ${\sigma_x}^2$ is the average of the squared deviation $(x - \bar{x})^2$. Thus exactly the argument leading to (5.15) gives, after many trials,

$${\sigma_x}^2 = \int_{-\infty}^{\infty} (x - \bar{x})^2 f(x)\,dx. \tag{5.16}$$

5.3. *The Normal Distribution*

Different types of measurements have different limiting distributions. Not all limiting distributions have the symmetrical bell shape illustrated in Section 5.2. (For example, the binomial and Poisson distributions discussed in Chapters 10 and 11 are usually not symmetrical.) Nevertheless, a great many measurements are found to have a symmetrical bell-shaped curve for their limiting distribution. In fact, we shall prove in

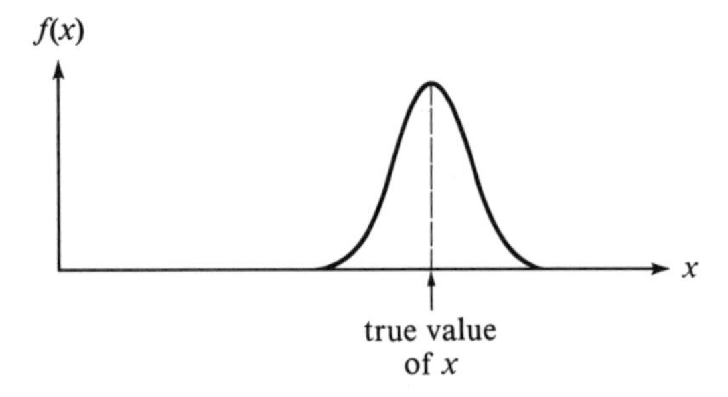

Figure 5.7. The limiting distribution for a measurement that is subject to many small random errors. The distribution is bell-shaped and centered on the true value of the measured quantity x.

Chapter 10 that if a measurement is subject to many small sources of random error and negligible systematic error, then the measured values will be distributed on a bell-shaped curve, and that this curve will be centered on the true value of x, as in Figure 5.7. In the remainder of this chapter we shall confine attention to measurements with this property.

If our measurements have appreciable systematic errors, then we would *not* expect the limiting distribution to be centered on the true value. Random errors are equally likely to push our readings above or below the true value. If all errors are random, after many measurements there will be as many observations above the true value as there are below it, and our distribution of results will therefore be centered on the true value. But a systematic error (like a tape measure that is stretched, or a clock that runs slow) pushes all values in one direction, and so pushes the distribution of observed values off-center from the true value. In this chapter we are going to assume that the distribution is centered on the true value. This is equivalent to assuming that all systematic errors have been reduced to a negligible level.

We must now address briefly a question that we have avoided discussing so far: What is the "true value" of a physical quantity? This is a hard question, to which there is no satisfactory, simple answer. Since it is obvious that no measurement can exactly determine the true value of any continuous variable (a length, a time, etc.), it is not even clear that the true value of such a quantity exists. Nevertheless, it will be very convenient to assume that every physical quantity does have a true value; and we will make this assumption.

We can think of the true value of a quantity as that value to which one approaches closer and closer as one makes more and more measurements, more and more carefully. As such, the "true value" is an idealization, similar to the mathematician's point with no size or line with no width; and like both of these it is a useful idealization. We will often denote the true values of measured quantities $x, y, \ldots$, by the corresponding capital letters $X, Y, \ldots$. If the measurements of x are subject to many small random errors, but negligible systematic errors, then their distribution will be a symmetrical bell-shaped curve, centered on the true value X.

The mathematical function that describes the bell-shaped curve is called the *normal distribution* or *Gauss function.*[2] The prototype of this function is

$$e^{-x^2/2\sigma^2}, \tag{5.17}$$

where σ is a fixed parameter that we will call the *width parameter*. The reader must become familiar with the properties of this function.

When $x = 0$, the Gauss function (5.17) is equal to one. The function is symmetric about $x = 0$, since it has the same value for x and $-x$. As x moves away from zero in either direction, $x^2/2\sigma^2$ increases, quickly if σ is small, more slowly if σ is large. Therefore, as x moves away from the origin, the function (5.17) decreases towards zero. Thus the general appearance of the Gauss function (5.17) is as shown in Figure 5.8. The picture explains the name "width parameter" for σ, since the bell shape is wide if σ is large, and narrow if σ is small.

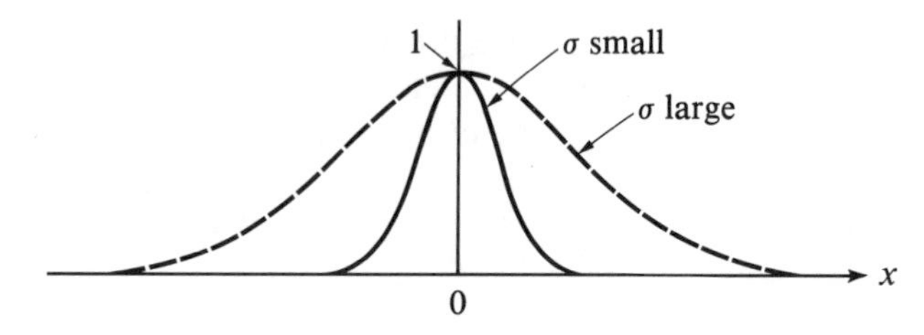

Figure 5.8. The Gauss function (5.17) is bell-shaped and centered on $x = 0$. The curve is wide if the width parameter σ is large, and narrow if σ is small.

The Gauss function (5.17) is a bell-shaped curve centered on $x = 0$. To obtain a bell-shaped curve centered on some other point $x = X$, we merely replace x in (5.17) by $x - X$. Thus the function

$$e^{-(x-X)^2/2\sigma^2} \tag{5.18}$$

[2] Other common names for the Gauss function are: the Gaussian function (or just "Gaussian"), the normal density function, and the normal error function. The last of these names is rather unfortunate, since the name "error function" is often used for the integral of the Gauss function (as we will discuss in Section 5.4).

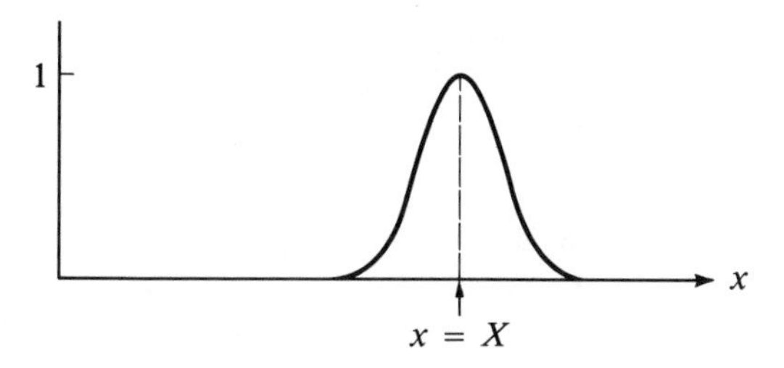

Figure 5.9. The Gauss function (5.18) is bell-shaped and centered on $x = X$.

has its maximum at $x = X$ and falls off symmetrically on either side of $x = X$, as in Figure 5.9.

The function (5.18) is not quite in its final form to describe a limiting distribution, because any distribution must be *normalized*, that is, must satisfy

$$\int_{-\infty}^{\infty} f(x)\,dx = 1. \tag{5.19}$$

To arrange this, we set

$$f(x) = Ne^{-(x-X)^2/2\sigma^2}. \tag{5.20}$$

(Multiplication by the factor N does not change the shape, nor does it shift the maximum at $x = X$.) We must then choose the "normalization factor" N so that $f(x)$ is normalized as in (5.19). This involves some elementary manipulation of integrals, which we give in some detail:

$$\int_{-\infty}^{\infty} f(x)\,dx = \int_{-\infty}^{\infty} Ne^{-(x-X)^2/2\sigma^2}\,dx. \tag{5.21}$$

To evaluate this kind of integral, it is always a good idea to change variables to simplify the integral. Thus we can set $x - X = y$ (in which case $dx = dy$) and get

$$= N\int_{-\infty}^{\infty} e^{-y^2/2\sigma^2}\,dy. \tag{5.22}$$

Next we can set $y/\sigma = z$ (in which case $dy = \sigma\,dz$), and get

$$= N\sigma\int_{-\infty}^{\infty} e^{-z^2/2}\,dz. \tag{5.23}$$

The remaining integral is one of the standard integrals of mathematical physics. It can be evaluated by elementary methods, but the details are not especially illuminating; so we will simply quote the result;[3]

$$\int_{-\infty}^{\infty} e^{-z^2/2}\, dz = \sqrt{2\pi}. \tag{5.24}$$

Returning to (5.21) and (5.23), we find that

$$\int_{-\infty}^{\infty} f(x)\, dx = N\sigma\sqrt{2\pi}.$$

Since this integral has to be one, we must choose the normalization factor N to be $N = 1/(\sigma\sqrt{2\pi})$.

We conclude that the correctly normalized Gauss, or normal, distribution function is

$$f_{X,\sigma}(x) = \frac{1}{\sigma\sqrt{2\pi}} e^{-(x-X)^2/2\sigma^2}. \tag{5.25}$$

Notice that we have added subscripts X, σ to indicate the center and width of the distribution. The function $f_{X,\sigma}(x)$ describes the limiting distribution of results in a measurement of a quantity x whose true value is X, if the measurement is subject only to random errors. Any measurements whose limiting distribution is given by the Gauss function (5.25) are said to be *normally distributed.*

We will explore the significance of the width parameter σ shortly. It is already clear that a small value of σ gives a sharply peaked distribution, corresponding to a precise measurement; whereas a large value of σ gives a broad distribution, corresponding to a low-precision measurement. In Figure 5.10 we show two examples of Gauss distributions with different centers X and widths σ. Note how the factor σ in the denominator of (5.25) guarantees that a narrower distribution (σ smaller) is automatically higher at its center, as it must be in order that the total area under the curve be 1.

We saw in Section 5.2 that knowledge of the limiting distribution for a measurement lets us compute the average value $\bar{x}$ expected after a very large number of trials. According to (5.15) the expected average after many trials is

$$\bar{x} = \int_{-\infty}^{\infty} x f(x)\, dx. \tag{5.26}$$

[3] For a derivation, see, for example, Hugh D. Young *Statistical Treatment of Experimental Data* (McGraw-Hill, 1962), Appendix D.

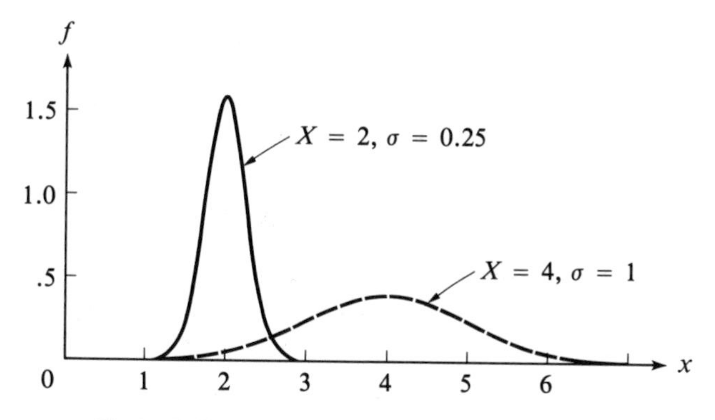

Figure 5.10. Two normal, or Gauss, distributions.

If the limiting distribution is the Gauss distribution $f_{X,\sigma}(x)$ centered on the true value X, then this integral can be evaluated. Before we do this we should note that it is almost obvious that the average $\bar{x}$ after very many trials will be X, because the symmetry of the Gauss function about X implies that as many results will fall any distance above X as will fall an equal distance below X. Thus the average should be X.

We can calculate the integral (5.26) for the Gauss distribution as follows:

$$\bar{x} = \int_{-\infty}^{\infty} x f_{X,\sigma}(x)\,dx$$
$$= \frac{1}{\sigma\sqrt{2\pi}} \int_{-\infty}^{\infty} x e^{-(x-X)^2/2\sigma^2}\,dx. \tag{5.27}$$

If we make the change of variables $y = x - X$, then $dx = dy$ and $x = y + X$. Thus the integral (5.27) becomes two terms,

$$\bar{x} = \frac{1}{\sigma\sqrt{2\pi}}\left(\int_{-\infty}^{\infty} y e^{-y^2/2\sigma^2}\,dy + X\int_{-\infty}^{\infty} e^{-y^2/2\sigma^2}\,dy\right). \tag{5.28}$$

The first integral here is exactly zero, since the contribution from any point y is exactly canceled by that from the point $-y$. The second integral is the normalization integral encountered in (5.22), and has the value $\sigma\sqrt{2\pi}$. This cancels with the $\sigma\sqrt{2\pi}$ in the denominator and leaves the expected answer, that

$$\bar{x} = X, \tag{5.29}$$

after many trials. In other words, if the measurements are distributed according to the Gauss distribution $f_{X,\sigma}(x)$, then, after many, many trials, the mean value $\bar{x}$ is the true value X, on which the Gauss function is centered.

The result (5.29) would be exactly true only if we could make infinitely many measurements. Its practical usefulness is that if we make a large (but finite) number of trials, then our average will be *close* to X.

Another interesting quantity to compute is the *standard deviation* σ_x after a large number of trials. According to (5.16), this is given by

$$\sigma_x{}^2 = \int_{-\infty}^{\infty} (x - \bar{x})^2 f_{X,\sigma}(x)\, dx. \tag{5.30}$$

This integral is easily evaluated. We replace $\bar{x}$ by X, make the substitutions $x - X = y$ and $y/\sigma = z$, and finally integrate by parts to obtain the result (see Problem 5.6)

$$\sigma_x{}^2 = \sigma^2, \tag{5.31}$$

after many trials. In other words, the width parameter σ of the Gauss function $f_{X,\sigma}(x)$ is just the standard deviation that we would obtain after making many measurements. This is, of course, the reason why the letter σ is used for the width parameter, and explains why σ is often called the standard deviation of the Gauss distribution $f_{X,\sigma}(x)$. However, strictly speaking, σ is the standard deviation expected only after *infinitely many* trials. If we make some finite number of measurements (10 or 20, say) of x, then the observed standard deviation should be some approximation to σ, but we have no reason to think it will be *exactly* σ. In Section 5.5 we will consider further what can be said about the mean and standard deviation after we have made some reasonable finite number of trials.

5.4. *The Standard Deviation as 68 Percent Confidence Limit*

The limiting distribution $f(x)$ for measurement of some quantity x tells us the probability of obtaining any given value of x. Specifically, the integral

$$\int_a^b f(x)\, dx$$

is the probability that any one measurement gives an answer in the range $a \leqslant x \leqslant b$. If the limiting distribution is the Gauss function $f_{X,\sigma}(x)$, then this integral can be evaluated. In particular, we can now calculate the

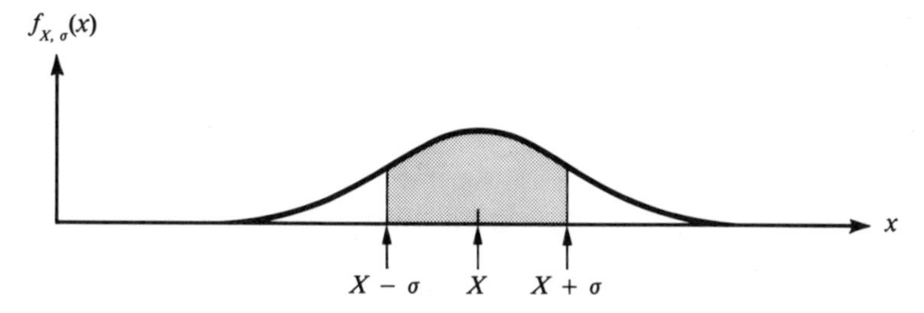

Figure 5.11. The shaded area between $X \pm \sigma$ is the probability of a measurement within one standard deviation of X.

probability (discussed in Chapter 4) that a measurement will fall within one standard deviation σ of the true value X. This probability is

$$P(\text{within } \sigma) = \int_{X-\sigma}^{X+\sigma} f_{X,\sigma}(x)\,dx \tag{5.32}$$

$$= \frac{1}{\sigma\sqrt{2\pi}} \int_{X-\sigma}^{X+\sigma} e^{-(x-X)^2/2\sigma^2}\,dx. \tag{5.33}$$

This integral is illustrated in Figure 5.11. It can be simplified in the now familiar way by substituting $(x - X)/\sigma = z$. With this substitution, $dx = \sigma\,dz$ and the limits of integration become $z = \pm 1$. Therefore

$$P(\text{within } \sigma) = \frac{1}{\sqrt{2\pi}} \int_{-1}^{1} e^{-z^2/2}\,dz. \tag{5.34}$$

Before discussing the integral (5.34), we remark that we could equally have found the probability for an answer within 2σ of X, or within 1.5σ of X. More generally, we could calculate $P(\text{within } t\sigma)$, which means "the probability for an answer within $t\sigma$ of X," where t is any positive number. This probability is given by the area in Figure 5.12, and a calculation

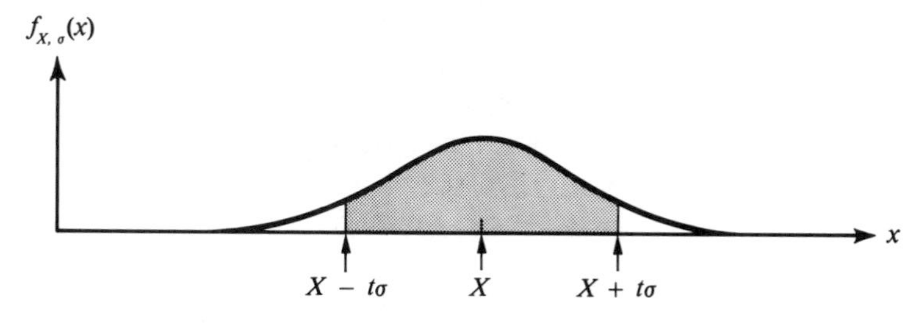

Figure 5.12. The shaded area between $X \pm t\sigma$ is the probability of a measurement within t standard deviations of X.

identical to that leading to (5.34) gives (Problem 5.7)

$$P(\text{within } t\sigma) = \frac{1}{\sqrt{2\pi}} \int_{-t}^{t} e^{-z^2/2}\, dz. \tag{5.35}$$

The integral (5.35) is a standard integral of mathematical physics, and is often called the *error function*, denoted erf(t), or the *normal error integral.* It cannot be evaluated analytically, but is easily calculated with a pocket calculator. It is plotted as a function of t, and a few values tabulated, in Figure 5.13. A more complete tabulation can be found in Appendix A at the end of the book. (See also Appendix B, which shows a different, but closely related, integral.)

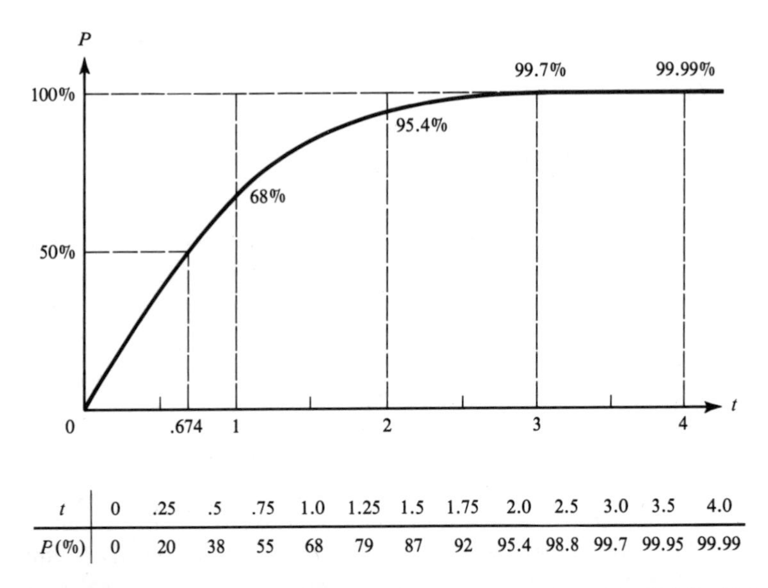

t	0	.25	.5	.75	1.0	1.25	1.5	1.75	2.0	2.5	3.0	3.5	4.0
P(%)	0	20	38	55	68	79	87	92	95.4	98.8	99.7	99.95	99.99

Figure 5.13. The probability $P(\text{within } t\sigma)$ that a measurement of x will fall within t standard deviations of the true value $x = X$. Two common names for this function are the *normal error integral* and the *error function*, erf(t).

We first note from Figure 5.13 that the probability that a measurement will fall within one standard deviation of the true answer is 68 percent, as was anticipated in Chapter 4 (where it was said to be "about 70 percent"). If we quote the standard deviation as our uncertainty in such a measurement (i.e., write $x = x_{\text{best}} \pm \delta x$, and take $\delta x = \sigma$), then we can be 68 percent confident that we are within σ of the correct answer.

We can also see in Figure 5.13 that the probability $P(\text{within } t\sigma)$ rapidly approaches 100 percent as t increases. The probability that a measurement will fall inside 2σ is 95.4 percent; that for 3σ is 99.7 percent. To put these results another way, the probability that a measurement will fall *outside* one standard deviation is quite appreciable (32 percent), that it will lie outside 2σ is much smaller (4.6 percent), and that it will lie outside 3σ is extremely small (.3 percent).

There is, of course, nothing sacred about the number 68 percent; it just happens to be the confidence associated with the standard deviation σ. One alternative to the standard deviation is the so-called *probable error*, P.E., which is defined as that distance for which there is a 50 percent probability of a measurement between $X \pm$ P.E. It can be seen from Figure 5.13 that (for a measurement that is normally distributed) the probable error is

$$\text{P.E.} \approx .67\sigma.$$

Some experimenters like to quote the probable error, P.E., as the uncertainty in their measurements. Nonetheless, the standard deviation σ is the most popular choice, since its properties are so simple.

5.5. *Justification of the Mean as Best Estimate*

For the last three sections we have been discussing the *limiting distribution* $f(x)$, the distribution one would obtain from an infinite number of measurements of a quantity x. If $f(x)$ were known, then we could calculate the mean $\bar{x}$ and standard deviation σ obtained after infinitely many measurements, and (at least for the normal distribution) we would also know the true value, X. Unfortunately, we never do know the limiting distribution. In practice we have some finite number of measured values (5, or 10, or perhaps 50),

$$x_1, x_2, \ldots, x_N,$$

and our problem is to arrive at *best estimates* of X and σ, based on these N measured values.

If the measurements follow a normal distribution $f_{X,\sigma}(x)$ and if we knew the parameters X and σ, then we could calculate the probability of obtaining the values $x_1, \ldots, x_N$ that were actually obtained. Thus, the probability of getting a reading near x_1, in a small interval dx_1, is

$$P(x \text{ between } x_1 \text{ and } x_1 + dx_1) = \frac{1}{\sigma\sqrt{2\pi}}\, e^{-(x_1 - X)^2/2\sigma^2}\, dx_1. \qquad (5.36)$$

In practice we are not interested in the size of the interval dx_1 (nor the factor $\sqrt{2\pi}$); so we abbreviate this to

$$P(x_1) \propto \frac{1}{\sigma} e^{-(x_1 - X)^2/2\sigma^2}. \tag{5.37}$$

We shall refer to (5.37) as the probability of getting the value x_1, although strictly speaking it is the probability of getting a value in an interval near x_1 as in (5.36).

The probability of obtaining the second reading x_2 is

$$P(x_2) \propto \frac{1}{\sigma} e^{-(x_2 - X)^2/2\sigma^2}, \tag{5.38}$$

and we can similarly write down all the probabilities ending with

$$P(x_N) \propto \frac{1}{\sigma} e^{-(x_N - X)^2/2\sigma^2}. \tag{5.39}$$

Equations (5.37) to (5.39) give the probabilities of obtaining each of the readings $x_1, \ldots, x_N$, calculated in terms of the assumed limiting distribution $f_{X,\sigma}(x)$. The probability that we observe the whole set of N readings is just the product of these separate probabilities,[4]

$$P_{X,\sigma}(x_1, \ldots, x_N) = P(x_1) \times P(x_2) \times \cdots \times P(x_N)$$

or

$$P_{X,\sigma}(x_1, \ldots, x_N) \propto \frac{1}{\sigma^N} e^{-\Sigma(x_i - X)^2/2\sigma^2}. \tag{5.40}$$

It is most important to understand the significance of the various quantities in (5.40). The numbers $x_1, \ldots, x_N$ are the actual results of N measurements; thus $x_1, \ldots, x_N$ are known, fixed numbers. The quantity $P_{X,\sigma}(x_1, \ldots, x_N)$ is the probability for obtaining the N results $x_1, \ldots, x_N$, calculated in terms of X and σ, the true value of x and the width parameter of its distribution. The numbers X and σ are *not* known; we want to find best estimates for X and σ based on the given observations $x_1, \ldots, x_N$. We have added subscripts X and σ to the probability (5.40) to emphasize that it depends on the (unknown) values of X and σ.

[4] We are using the well-known result that the probability for several independent events is the product of their separate probabilities. For example, the probability of throwing a "heads" with a coin is $\frac{1}{2}$, and that of throwing a "six" with a die is $\frac{1}{6}$. Therefore, the probability of throwing a "heads" *and* a "six" is $(\frac{1}{2}) \times (\frac{1}{6}) = \frac{1}{12}$.

Since the actual values of X and σ are unknown, we might imagine guessing values X', σ' and then using those guessed values to compute the probability $P_{X',\sigma'}(x_1, \ldots, x_N)$. If we next guessed two new values, X'' and σ'', and found that the corresponding probability $P_{X'',\sigma''}(x_1,\ldots,x_N)$ was larger, then we would naturally regard the new values, X'' and σ'', as better estimates for X and σ. Continuing in this way, we could imagine hunting for the values of X and σ that make $P_{X,\sigma}(x_1, \ldots, x_N)$ as large as possible, and these values would be regarded as the best estimates for X and σ.

This plausible procedure for finding the best estimates for X and σ is called by statisticians the *principle of maximum likelihood*. It can be stated briefly as follows.

Given the N observed measurements $x_1, \ldots, x_N$, the best estimates for X and σ are those values for which the observed $x_1, \ldots, x_N$ are most likely. That is, the best estimates for X and σ are those values for which $P_{X,\sigma}(x_1, \ldots, x_N)$ is maximum, given that here

$$P_{X,\sigma}(x_1, \ldots, x_N) \propto \frac{1}{\sigma^N} e^{-\Sigma(x_i - X)^2/2\sigma^2}. \tag{5.41}$$

Using this principle, we can easily find the best estimate for the true value X. Obviously (5.41) is *maximum* if the sum in the exponent is *minimum*. Thus the best estimate for X is that value of X for which

$$\sum_{i=1}^{N} (x_i - X)^2/\sigma^2 \tag{5.42}$$

is minimum. To locate this minimum, we differentiate with respect to X and set the derivative equal to zero, giving

$$\sum_{i=1}^{N} (x_i - X) = 0$$

or

$$X = \frac{\sum x_i}{N} \qquad \text{(best estimate).} \tag{5.43}$$

That is, the *best estimate* for the true value X is the mean of our N measurements, $\bar{x} = \sum x_i/N$, a result we have been assuming without proof since Chapter 1.

To find the best estimate for σ, the width of the limiting distribution, is a little harder, since the probability (5.41) is a more complicated function of σ. We must differentiate (5.41) with respect to σ, and set the derivative equal to zero. (We leave the details to the reader; see Problem 5.10.) This gives the value of σ that maximizes (5.41), and that is therefore the best estimate for σ, as

$$\sigma = \sqrt{\frac{1}{N}\sum_{i=1}^{N}(x_i - X)^2} \qquad \text{(best estimate).} \tag{5.44}$$

The true value X is unknown. Thus, in practice, we have to replace X in (5.44) by our best estimate for X, namely the mean $\bar{x}$. This yields the estimate

$$\sigma = \sqrt{\frac{1}{N}\sum_{i=1}^{N}(x_i - \bar{x})^2}. \tag{5.45}$$

In other words, our estimate for the width σ of the limiting distribution is the standard deviation of the N observed values, $x_1, \ldots, x_N$, as originally defined in (4.6).

The reader may have been surprised that the estimate (5.45) is the same as our original definition (4.6), using N, of the standard deviation, instead of our "improved" definition, using $N - 1$. In fact, in passing from the best estimate (5.44) to the expression (5.45), we have glossed over a rather elegant subtlety. The best estimate (5.44) involves the true value X, whereas in (5.45) we have replaced X by $\bar{x}$ (our best estimate for X). Now, these numbers are generally not the same, and it is easily seen that the number (5.45) is *always less than*, or at most equal to, (5.44).[5] Thus in passing from (5.44) to (5.45) we have consistently *underestimated* the width σ. It is fairly easy to estimate the amount by which (5.45) is less than (5.44), though we shall not do so here. The result is that the best approximation for σ is not (5.45) itself, but is obtained by multiplying (5.45) by about $\sqrt{N/(N-1)}$. Thus our final conclusion is that the best estimate for the width σ is the "improved" standard deviation of the measured values $x_1, \ldots, x_N$,

$$\sigma = \sqrt{\frac{1}{N-1}\sum_{i=1}^{N}(x_i - \bar{x})^2} \qquad \text{(best estimate).} \tag{5.46}$$

It may be well to pause and review the rather complicated story that we have unfolded so far. First, if the measurements of x are subject only to random errors, then their limiting distribution is the Gauss function $f_{X,\sigma}(x)$, centered on the true value X and with width σ. The width σ is the

[5] If we regard (5.44) as a function of X, then we have just seen that this function is minimum at $X = \bar{x}$. Thus (5.45) is always less than or equal to (5.44).

68 percent confidence limit, in that there is a 68 percent probability that any measurement will fall within a distance σ of the true value X. In practice, neither X nor σ is known. Instead, we know our N measured values $x_1, \ldots, x_N$, where N is as large as our time and patience allowed us to make it. Based on these N measured values, our best estimate of the true value X has been shown to be the mean $\bar{x} = \sum x_i/N$, and our best estimate of the width σ is the standard deviation σ_x of $x_1, \ldots, x_N$ as defined in (5.46). In Section 5.7 we will discuss the *reliability* of $\bar{x}$ as the best estimate for X; in a similar way we could discuss the reliability of σ_x as the best estimate for σ, but we will not do so here.

All the results of the last two sections depend on the assumption that our measurements are normally distributed.[6] Although this is a reasonable assumption, it is one that is difficult to verify in practice, and is sometimes not exactly true. This being the case, we should emphasize that, even when the distribution of measurements is *not* normal, it is almost always *approximately* normal, and one can safely use the ideas of this chapter, at least as good approximations.

5.6. *Justification of Addition in Quadrature*

We can now return to a problem discussed in Chapter 3, the propagation of errors. We stated there, without formal proof, that when errors are random and independent, they can be combined in quadrature according to certain standard rules, either the "simple rules" in (3.16) and (3.18), or the general rule in (3.47), which includes the "simple" rules as special cases. We are now ready to justify this use of addition in quadrature.

The problem of error propagation arises when we measure one or more quantities $x, \ldots, z$, all with uncertainties, and then use our measured values to calculate some quantity $q(x, \ldots, z)$. The main question is, of course, to decide on the uncertainty in our answer for q. If the quantities $x, \ldots, z$ are subject only to random errors, then they will be normally distributed with width parameters[7] $\sigma_x, \ldots, \sigma_z$, which we take to be the uncertainties associated with any single measurement of the corresponding quantities. The question to be decided now is this: Knowing the distributions of measurements of $x, \ldots, z$, what can we say about the distribution of values of q? And in particular, what will be the width of the distribution of values of q?

[6] And that systematic errors have been reduced to a negligible level.

[7] When we are discussing several different measured quantities $x, \ldots, z$, we use subscripts $x, \ldots, z$ to distinguish the width parameters of their limiting distributions. Thus σ_x denotes the width of the Gauss distribution $f_{X,\sigma_x}(x)$ for the measurements of x, and so on.

Measured Quantity Plus Fixed Number

We begin by considering two very simple special cases. First, suppose that we measure a quantity x and proceed to calculate the quantity

$$q = x + A, \tag{5.47}$$

where A is some fixed number with no uncertainty (like $A = 1$ or π). Suppose that the measurements of x are normally distributed about the true value X, with width σ_x, as in Figure 5.14(a). Then the probability of obtaining any value x (in a small interval dx) is $f_{X,\sigma_x}(x)\,dx$ or

$$\text{(probability of obtaining value } x) \propto e^{-(x-X)^2/2\sigma_x^2}. \tag{5.48}$$

Our problem is to deduce the probability of obtaining any value q of the quantity defined by (5.47). Now, from (5.47) it is clear that $x = q - A$ and hence that

$$\text{(probability of obtaining value } q) = \text{(probability of obtaining } x = q - A).$$

The second probability is given by (5.48), and so

$$\begin{aligned}\text{(probability of obtaining value } q) &\propto e^{-[(q-A)-X]^2/2\sigma_x^2} \\ &= e^{-[q-(X+A)]^2/2\sigma_x^2}.\end{aligned} \tag{5.49}$$

The result (5.49) shows that the calculated values of q are normally distributed, and are centered on the value $X + A$, with width σ_x, as shown

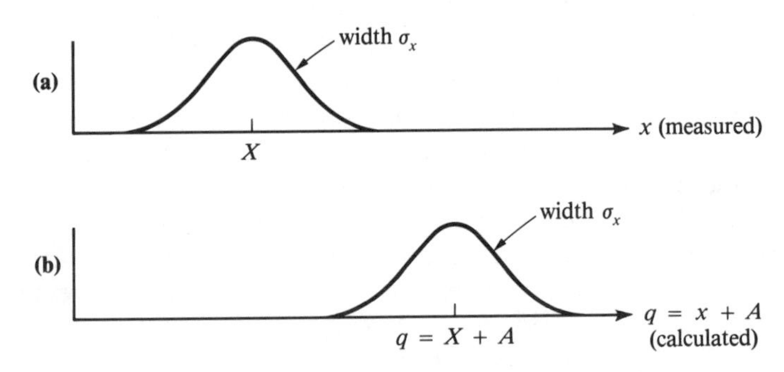

Figure 5.14. If the measured values of x are normally distributed with center $x = X$ and width σ_x, then the calculated values of $q = x + A$ (with A fixed and known) will be normally distributed with center $q = X + A$ and the same width σ_x.

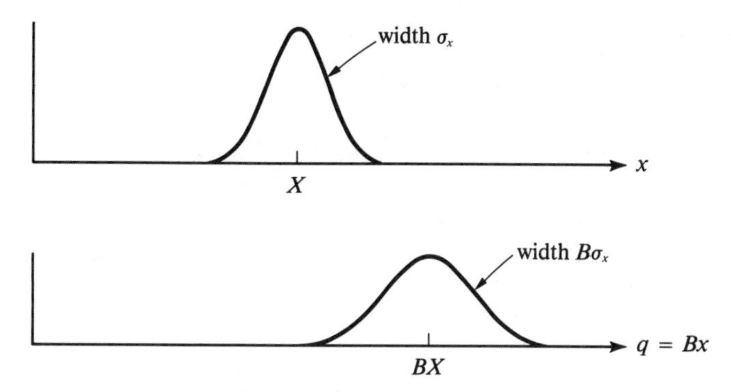

Figure 5.15. If the measured values of x are normally distributed with center $x = X$ and width σ_x, then the calculated values of $q = Bx$ (with B fixed and known) will be normally distributed with center BX and width $B\sigma_x$.

in Figure 5.14(b). In particular, the uncertainty in q is the same (namely, σ_x) as that in x, just as our rule in (3.16) would have predicted.

Measured Quantity Times Fixed Number

As a second simple example, suppose that we measure x and calculate the quantity

$$q = Bx,$$

where B is a fixed number (such as $B = 2$ or $B = \pi$). If the measurements of x are normally distributed, then, arguing exactly as before, we conclude that[8]

$$\begin{aligned}(\text{probability of obtaining value } q) &\propto (\text{probability of obtaining } x = q/B) \\ &\propto \exp\left[-\left(\frac{q}{B} - X\right)^2 \Big/ 2\sigma_x{}^2\right] \\ &= \exp[-(q - BX)^2/2B^2\sigma_x{}^2]. \qquad (5.50)\end{aligned}$$

In other words, the values of $q = Bx$ will be normally distributed, with center at $q = BX$ and with width $B\sigma_x$, as shown in Figure 5.15. In particular, the uncertainty in $q = Bx$ is B times that in x, just as our rule in (3.18) would have predicted.

[8] Here we introduce the alternative notation $\exp(z)$ for the exponential function, $\exp(z) \equiv e^z$. When the exponent z becomes complicated, the "exp" notation is more convenient to type or print.

Sum of Two Measured Quantities

As a first nontrivial example of error propagation, suppose that we measure two independent quantities x and y, and calculate their sum $x + y$. We suppose that the measurements of x and y are normally distributed about their true values X and Y, with widths σ_x and σ_y as in Figures 5.16(a) and (b), and we will try to find the distribution of the calculated values of $x + y$. What we will find is that the values of $x + y$ are normally distributed, that their center is the true value $X + Y$, and that the width of their distribution is

$$\sqrt{\sigma_x^2 + \sigma_y^2},$$

as in Figure 5.16(c). In particular, this justifies the rule of Chapter 3 that if x and y are subject to independent random uncertainties only, then the uncertainty in $x + y$ is the quadratic sum of the separate uncertainties in x and y.

To simplify our algebra, we assume at first that the true values X and Y are both zero. In this case the probability of getting any particular value of x is

$$P(x) \propto \exp\left(\frac{-x^2}{2\sigma_x^2}\right) \tag{5.51}$$

and that of y is

$$P(y) \propto \exp\left(\frac{-y^2}{2\sigma_y^2}\right). \tag{5.52}$$

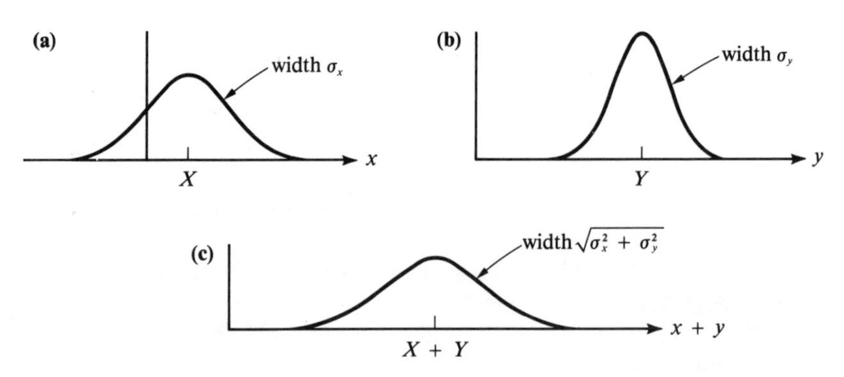

Figure 5.16. If the measurements of x and y are independent and normally distributed with centers X and Y and widths σ_x and σ_y, then the calculated values of $x + y$ are normally distributed with center $X + Y$ and width $\sqrt{\sigma_x^2 + \sigma_y^2}$.

Our problem now is to calculate the probability of obtaining any particular value of $x + y$. We first observe that, since x and y are independently measured, the probability of obtaining any given x *and* y is just the product of (5.51) and (5.52):

$$P(x, y) \propto \exp\left[-\frac{1}{2}\left(\frac{x^2}{\sigma_x{}^2} + \frac{y^2}{\sigma_y{}^2}\right)\right]. \tag{5.53}$$

Knowing the probability of obtaining any x *and* y, we can now calculate the probability for any given value of $x + y$. The first step is to rewrite the exponent in (5.53) in terms of the variable of interest, $x + y$. This can be done using the identity (which the reader can easily verify)

$$\frac{x^2}{A} + \frac{y^2}{B} = \frac{(x + y)^2}{A + B} + \frac{(Bx - Ay)^2}{AB(A + B)} \tag{5.54}$$

$$= \frac{(x + y)^2}{A + B} + z^2; \tag{5.55}$$

in the second line we have introduced the abbreviation z^2 for the second term on the right of (5.54), since its value does not interest us anyway.

If we substitute (5.55) into (5.53), replacing A with $\sigma_x{}^2$ and B with $\sigma_y{}^2$, we obtain

$$P(x, y) \propto \exp\left[-\frac{(x + y)^2}{2(\sigma_x{}^2 + \sigma_y{}^2)} - \frac{z^2}{2}\right]. \tag{5.56}$$

This probability for obtaining given values of x and y can just as well be viewed as the probability of obtaining given values of $x + y$ and z. Thus we can rewrite (5.56) as

$$P(x + y, z) \propto \exp\left[\frac{-(x + y)^2}{2(\sigma_x{}^2 + \sigma_y{}^2)}\right]\exp\left[\frac{-z^2}{2}\right]. \tag{5.57}$$

Finally, what we want is the probability of obtaining a given value of $x + y$, *irrespective of the value of* z. This is obtained by summing, or rather integrating, (5.57) over all possible values of z; that is,

$$P(x + y) = \int_{-\infty}^{\infty} P(x + y, z)\, dz. \tag{5.58}$$

When we integrate (5.57) with respect to z, the factor $\exp(-z^2/2)$ integrates to $\sqrt{2\pi}$, and we find

$$P(x + y) \propto \exp\left[\frac{-(x + y)^2}{2({\sigma_x}^2 + {\sigma_y}^2)}\right]. \tag{5.59}$$

This shows that the values of $x + y$ are normally distributed with width $\sqrt{{\sigma_x}^2 + {\sigma_y}^2}$ as anticipated.

Our proof is complete for the case when the true values of x and y are both zero, $X = Y = 0$. If X and Y are nonzero, we can proceed as follows: We first write

$$x + y = (x - X) + (y - Y) + (X + Y). \tag{5.60}$$

Here the first two terms are centered on zero, with widths σ_x and σ_y, by the result (5.49). Therefore the sum of the first two terms is normally distributed with width $\sqrt{{\sigma_x}^2 + {\sigma_y}^2}$. The third term in (5.60) is a fixed number. Therefore, by the result (5.49), it shifts the distribution to $(X + Y)$, but leaves the width unchanged. In other words, the values of $(x + y)$ as given by (5.60) are normally distributed about $(X + Y)$ with width $\sqrt{{\sigma_x}^2 + {\sigma_y}^2}$. This is the required result.

The General Case

Having justified the error-propagation formula for the special case of a sum $x + y$, we can justify the general formula for error propagation surprisingly simply. Suppose that we measure two independent quantities x and y whose observed values are normally distributed, and that we now calculate some quantity $q(x, y)$ in terms of x and y. The distribution of values of $q(x, y)$ is easily found by using the previous three results, as follows.

First, the widths σ_x and σ_y (the uncertainties in x and y) must be small. This means that we are concerned only with values of x close to X, and y close to Y, and we can use the approximation (3.42) to write

$$q(x, y) \approx q(X, Y) + \left(\frac{\partial q}{\partial x}\right)_{X,Y}(x - X) + \left(\frac{\partial q}{\partial y}\right)_{X,Y}(y - Y). \tag{5.61}$$

This approximation is good because the only values of x, y that occur significantly often are close to X, Y. We have given the two partial derivatives subscripts X, Y to emphasize that the derivatives are evaluated at X, Y and are therefore fixed numbers.

The approximation (5.61) expresses the desired quantity $q(x, y)$ as the sum of three terms. The first term $q(X, Y)$ is a fixed number; so it merely shifts the distribution of answers. The second term is the fixed number $\partial q/\partial x$ times $(x - X)$, whose distribution has width σ_x; so the values of the second term are centered on zero, with width

$$\left(\frac{\partial q}{\partial x}\right)\sigma_x.$$

Similarly, the values of the third term are centered on zero with width

$$\left(\frac{\partial q}{\partial y}\right)\sigma_y.$$

Combining the three terms in (5.61) and invoking the results already established, we conclude that the values of $q(x, y)$ are normally distributed about the true value $q(X, Y)$ with width

$$\sqrt{\left(\frac{\partial q}{\partial x}\sigma_x\right)^2 + \left(\frac{\partial q}{\partial y}\sigma_y\right)^2}. \tag{5.62}$$

If we identify the standard deviations σ_x and σ_y as the uncertainties in x and y, then the result (5.62) is precisely the rule in (3.47) for propagation of random errors, for the case when q is a function of just two variables, $q(x, y)$. If q depends on several variables, $q(x, y, \ldots, z)$, then the preceding argument can be immediately extended to establish the general rule in (3.47) for functions of several variables. Since all the rules of Chapter 3 (concerning propagation of random errors) can be derived from (3.47), all these rules are now justified.

5.7. *Standard Deviation of the Mean*

One more important result, quoted in Chapter 4, remains to be proved. This concerns the standard deviation of the mean $\sigma_{\bar{x}}$. We have proved (in Section 5.5) that if we make N measurements $x_1, \ldots, x_N$ of a quantity x (which is normally distributed), then the best estimate of the true value X is the mean $\bar{x}$ of $x_1, \ldots, x_N$. In Chapter 4 we stated that the uncertainty in this estimate is the standard deviation of the mean,

$$\sigma_{\bar{x}} = \sigma_x/\sqrt{N}. \tag{5.63}$$

We are now ready to prove this result. The proof is so surprisingly brief that you will need to follow it very carefully.

We suppose that the measurements of x are normally distributed about the true value X with width parameter σ_x. We now want to know the reliability of *the average of the N measurements.* To analyze this, we naturally imagine repeating our N measurements many times; that is, we imagine performing a whole sequence of experiments, in each of which we make N measurements and compute the average. We now wish to find out the distribution of these many determinations of the average of N measurements. And this is easily done.

In each experiment we measure N quantities $x_1, \ldots, x_N$, and then compute the function

$$\bar{x} = \frac{x_1 + \cdots + x_N}{N}. \tag{5.64}$$

The calculated quantity ($\bar{x}$) is a simple function of the measured quantities $x_1, \ldots, x_N$, and we can easily find the distribution of our answers for $\bar{x}$ by using propagation of errors. The only unusual feature of the function (5.64) is that all the measurements $x_1, \ldots, x_N$ happen to be measurements of the same quantity, with the same true value X and the same width σ_x.

We first observe that, since each of the measured quantities $x_1, \ldots, x_N$ is normally distributed, the same is true for the function $\bar{x}$ given by (5.64). Second, the true value for each of $x_1, \ldots, x_N$ is X; so the true value of $\bar{x}$ as given by (5.64) is

$$\frac{X + \cdots + X}{N} = X.$$

Thus, after making many determinations of the average $\bar{x}$ of N measurements, our many results for $\bar{x}$ will be normally distributed about the true value X. The only remaining question (and the most important question) is to find the width of our distribution of answers. According to (5.62), rewritten for N variables, this width is

$$\sigma_{\bar{x}} = \sqrt{\left(\frac{\partial \bar{x}}{\partial x_1}\sigma_{x_1}\right)^2 + \cdots + \left(\frac{\partial \bar{x}}{\partial x_N}\sigma_{x_N}\right)^2}. \tag{5.65}$$

Since $x_1, \ldots, x_N$ are all measurements of the same quantity x, their widths are all the same and are all equal to σ_x,

$$\sigma_{x_1} = \cdots = \sigma_{x_N} = \sigma_x$$

We also see from (5.64) that all the partial derivatives in (5.65) are the same:

$$\frac{\partial \bar{x}}{\partial x_1} = \cdots = \frac{\partial \bar{x}}{\partial x_N} = \frac{1}{N}.$$

Therefore (5.65) reduces to

$$\begin{aligned} \sigma_{\bar{x}} &= \sqrt{\left(\frac{1}{N}\sigma_x\right)^2 + \cdots + \left(\frac{1}{N}\sigma_x\right)^2} \\ &= \sqrt{N\frac{\sigma_x^2}{N^2}} \\ &= \sigma_x/\sqrt{N}, \end{aligned} \tag{5.66}$$

as required.

We have arrived at the desired result (5.66) so quickly that we probably need to pause and review what it signifies. We imagined a large number of experiments, in each of which we made N measurements of x and then computed the average $\bar{x}$ of those N measurements. We have shown that, after repeating this experiment many times, our many answers for $\bar{x}$ will be normally distributed, that they will be centered on the true value X, and that the width of their distribution is $\sigma_{\bar{x}} = \sigma_x/\sqrt{N}$, as shown in Figure 5.17 for $N = 10$. This width $\sigma_{\bar{x}}$ is the 68 percent confidence limit for our

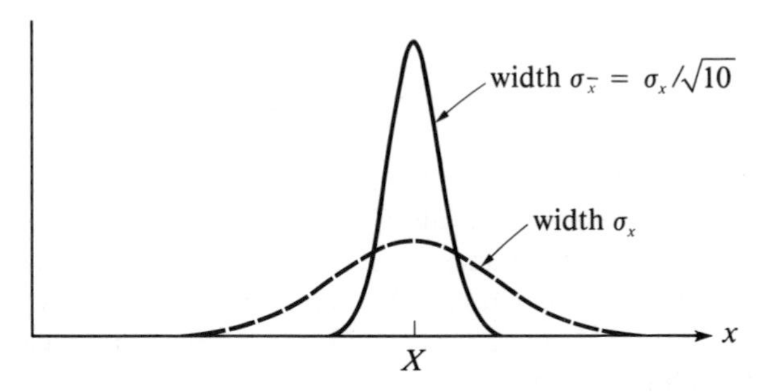

Figure 5.17. The individual measurements of x are normally distributed about X with width σ_x (dashed curve). If we use the same equipment to make many determinations of the average of 10 measurements, then the results $\bar{x}$ will be normally distributed about X with width $\sigma_{\bar{x}} = \sigma_x/\sqrt{10}$ (solid curve).

experiment. If we find the mean of N measurements *once*, then we can be 68 percent confident that our answer lies within a distance $\sigma_{\bar{x}}$ of the true value X. This is exactly what we would like to signify by the *uncertainty in the mean.* It also explains clearly why that uncertainty is called the standard deviation of the mean.

With this simple and elegant proof, we have now justified all the results quoted in earlier chapters concerning random uncertainties.

5.8. Confidence

We can now return to two questions first raised, but not completely answered, in Chapter 2. First, what do we mean by the now familiar statement that we are "reasonably confident" that some measured quantity lies in the range $x_{\text{best}} \pm \delta x$? And more generally, how can we give a quantitative value to our confidence in any experimental result?

For the first question, the answer should by now be clear. If we measure a quantity x several times (as one usually would), then our best estimate for x is the mean $\bar{x}$, and its standard deviation, $\sigma_{\bar{x}}$, is our best measure of its uncertainty. We would report a conclusion that

$$(\text{value of } x) = \bar{x} \pm \sigma_{\bar{x}},$$

meaning that, from our observations, we expect 68 percent of any subsequent measurements of x, made equally carefully, to fall in the range $\bar{x} \pm \sigma_{\bar{x}}$.

We could choose to characterize our uncertainty differently. For example, we might choose to state our conclusion as

$$(\text{value of } x) = \bar{x} \pm 2\sigma_{\bar{x}};$$

here we would be stating the range in which we expect 95 percent of all comparable measurements to fall. Clearly the essential point in stating any measured value is to state a range (or uncertainty) and *the confidence level corresponding to that range.* The commonest choice is to give the standard deviation of our answer, with its familiar significance as the 68 percent confidence limit.

As was emphasized in Chapter 2, almost all experimental conclusions involve the comparison of two or more numbers. With our statistical theory, we can now give a quantitative significance to many such comparisons. Here we will consider just one type of experiment, one in which we arrive at a number and compare our result with some known, expected answer. Notice that this general description fits many interesting experiments. For instance, in an experiment to check conservation of momentum, we may measure initial and final momenta, p and p', to verify that $p = p'$ (within uncertainties), but we can equally well regard this as finding a value for $(p - p')$ to be compared with the expected answer of zero. More generally, when we wish to compare any two measurements that are supposedly the same, we can form their difference and compare it with the expected answer of zero. Any experiment where one measures a quantity (like g, the acceleration of gravity) for which an accurate accepted value is known is also of this type, with the expected answer being the known accepted value.

Let us suppose that a student measures some quantity x (like the difference of two momenta that are supposedly equal) in the form

$$(\text{value of } x) = x_{\text{best}} \pm \sigma,$$

where σ denotes the standard deviation of his answer (which would be the SDOM if x_{best} was the mean of several measurements). He now wishes to compare this answer with the expected answer x_{exp}.

In Chapter 2 we argued that if the discrepancy $|x_{\text{best}} - x_{\text{exp}}|$ is less than (or only slightly more than) σ, then the agreement is satisfactory, but if $|x_{\text{best}} - x_{\text{exp}}|$ is much greater than σ, it is not satisfactory. As far as they go, these criteria are correct; but they give no quantitative measure of how good or bad the agreement is. Nor do they tell us where to draw the boundary of acceptability. Would a discrepancy of 1.5σ have been satisfactory? Or 2σ?

We can now answer these questions if we assume that our student's measurement was governed by a normal distribution (as is certainly reasonable). We start by making two working hypotheses about this distribution:

(a) the distribution is centered on the expected answer x_{exp};
(b) the width parameter of the distribution is equal to the student's estimate σ.

Hypothesis (a) is, of course, what the student hopes is true. It amounts to assuming that all systematic errors had been reduced to a negligible level (so that the distribution was centered on the true value) and that the true value was indeed x_{exp} (i.e., that the reasons for expecting x_{exp} were correct). Hypothesis (b) is an approximation, since σ must have been an estimate of the standard deviation, but it is a reasonable approximation if the number of measurements on which σ is based is large.[9] Taken together, our two hypotheses amount to assuming that the student's procedures and calculations were essentially correct.

We must now decide whether the student's value x_{best} was a reasonable one to obtain if our hypotheses were correct. If the answer is yes, then there is no reason to doubt the hypotheses, and all is well; if the answer is no, then the hypotheses must be doubted, and the student must examine the possibilities of mistakes in the measurements or calculations, of undetected systematic errors, and of the expected answer x_{exp} being incorrect.

We first determine the discrepancy, $|x_{best} - x_{exp}|$, and then

$$t = \frac{|x_{best} - x_{exp}|}{\sigma}, \tag{5.67}$$

the number of standard deviations by which x_{best} differs from x_{exp}. Next, from the table of the normal error integral in Appendix A, we can find the probability (given our hypotheses) of obtaining an answer that differs from x_{exp} by t or more standard deviations. This is

$$P(\text{outside } t\sigma) = 1 - P(\text{within } t\sigma). \tag{5.68}$$

If this probability is large, then the discrepancy $|x_{best} - x_{exp}|$ is perfectly reasonable, and the result x_{best} is acceptable; if the probability in (5.68) is found to be "unreasonably small," then the discrepancy must be judged *significant* (i.e., unacceptable), and our unlucky student must try to find out what has gone wrong.

Suppose, for example, the discrepancy $|x_{best} - x_{exp}|$ is one standard deviation. The probability of a discrepancy this large or larger is the familiar 32 percent. Clearly a discrepancy of one standard deviation is quite likely to occur and is, therefore, insignificant. At the opposite extreme, the probability $P(\text{outside } 3\sigma)$ is just 0.3 percent, and, if our hypotheses are correct, it is most unlikely that we would get a discrepancy of 3σ. Turning this around, if our student's discrepancy *is* 3σ, then it is most unlikely that our hypotheses were correct.

[9] We are going to judge the reasonableness of our measurement, x_{best}, by comparing $|x_{best} - x_{exp}|$ with σ, our *estimate* of the width of the normal distribution concerned. If the number of measurements on which σ was based is *small*, then this estimate may be fairly unreliable, and the confidence levels will be correspondingly inaccurate (although still a useful rough guide). With a small number of measurements, the accurate calculation of confidence limits requires use of the so-called "Student's t distribution," which allows for the probable variations in our estimate σ of the width. See H. L. Alder and E. B. Roessler, *Introduction to Probability and Statistics* (W. H. Freeman, 6th ed., 1977) Chapter 10.

The boundary between acceptability and unacceptability depends on the level below which we judge that a discrepancy is unreasonably improbable. This level is a matter of opinion, to be decided by the experimenter. However, many people regard 5 percent as a fair boundary for "unreasonable improbability." If we accept this choice, then a discrepancy of 2σ would be just unacceptable, since $P(\text{outside } 2\sigma) = 4.6$ percent. In fact, from the table in Appendix A, we see that any discrepancy greater than 1.96σ is unacceptable at this 5 percent level. At the 2 percent level, any discrepancy greater than 2.32σ would be unacceptable, and so on.

We see that we still do not have a clear-cut answer that a certain measured value x_{best} is, or is not, acceptable. However, our theory of the normal distribution has given us a clear and quantitative measure of the reasonableness of any particular answer. And this is the best we are entitled to ask for.

There are, of course, more complicated kinds of experiments, the analysis of whose results is correspondingly more involved. However, most of the basic principles have already been illustrated by the simple and important case discussed here. The reader who wishes to pursue further examples can find several in Part II of this book.

Problems

Reminder: An asterisk indicates that the problem is discussed, or its answer given, in the Answers section at the back of the book.

***5.1** (Section 5.1). A student measures the angular momenta L_i and L_f of a rotating system before and after adding an extra mass. To check conservation of angular momentum, he calculates $L_i - L_f$ (expecting the answer 0). He repeats the measurement 50 times, and collects his answers into bins as shown in Table 5.3, which shows his results (in some

Table 5.3.

	Bin								
Number after	−9 to −7	−7 to −5	−5 to −3	−3 to −1	−1 to 1	1 to 3	3 to 5	5 to 7	7 to 9
5 trials	0	1	2	0	1	0	1	0	0
10 trials	0	1	2	2	3	1	1	0	0
50 trials	1	3	7	8	10	9	6	4	2

unspecified units) after 5, 10, and 50 trials. Draw bin histograms for each of these three cases. (Be careful to choose your scales so that the area of each rectangle is the fraction of events in the corresponding bin.)

***5.2** (Section 5.2). The limiting distribution of results in some hypothetical measurement has the form

$$f(x) = \begin{cases} C & \text{for } |x| < a, \\ 0 & \text{otherwise.} \end{cases}$$

(a) Use the normalization condition (5.13) to calculate the constant C in terms of a.
(b) Sketch the limiting distribution. What is the significance of the constant a?
(c) Using Equations (5.15) and (5.16), calculate the mean $\bar{x}$ and standard deviation that would be found after many measurements.

5.3 (Section 5.3). Using proper squared paper and clearly labeled axes, make good plots of the Gauss distribution

$$f_{X,\sigma}(x) = \frac{1}{\sigma\sqrt{2\pi}} e^{-(x-X)^2/2\sigma^2}$$

for $X = 2$, $\sigma = 1$, and for $X = 3$, $\sigma = .3$. Use your calculator to compute the values of $f_{X,\sigma}(x)$. If it has two memories to store $\sigma\sqrt{2\pi}$ and $-2\sigma^2$, this will speed your calculations. If you remember that the function is symmetrical about $x = X$, this halves the number of calculations needed. Put both graphs on the same plot for comparison.

***5.4** (Section 5.3). If you have not yet done so, plot the third histogram of Problem 5.1. The student of Problem 5.1 decides that the distribution of his results is consistent with the Gauss distribution $f_{X,\sigma}(x)$ centered on $X = 0$ with width $\sigma = 3.4$. Draw this distribution on the same graph, and compare with your histogram. (Read the hints to Problem 5.3. Note that you have no quantitative way to compare the fit; all you can do is see if the Gauss function *appears* to fit the histogram satisfactorily.)

5.5 (Section 5.3). The width of a Gauss distribution is usually characterized by the parameter σ. An alternative parameter with a simple geometrical interpretation is the *full width at half maximum*, or FWHM. This is the distance between the two points x where $f_{X,\sigma}(x)$ is half its maximum value, as in Figure 5.18. Prove that

$$\text{FWHM} = 2\sigma\sqrt{2\ln 2} = 2.35\sigma.$$

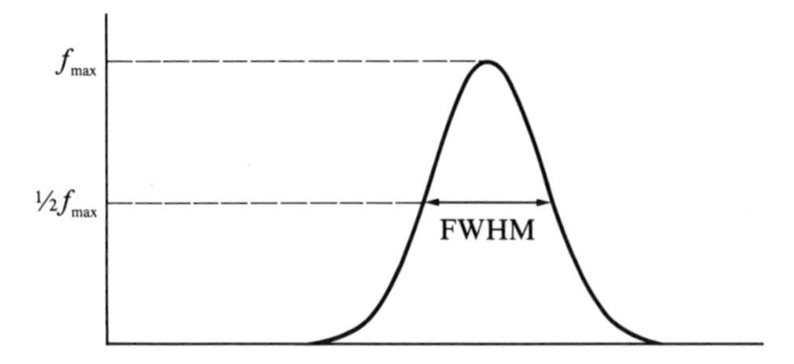

Figure 5.18. The full width at half maximum, FWHM.

This means that the half-maxima are at the points $X \pm 1.17\sigma$ or, very roughly, $X \pm \sigma$.

***5.6** (Section 5.3). Give in detail the steps leading from (5.30) to (5.31) to show that the standard deviation σ_x of very many measurements that are distributed normally with width parameter σ is $\sigma_x = \sigma$.

5.7 (Section 5.4). If the measurements of some quantity x are governed by the Gauss distribution $f_{X,\sigma}(x)$, then the probability of obtaining a value between $X - t\sigma$ and $X + t\sigma$ is obviously

$$P(\text{within } t\sigma) = \int_{X-t\sigma}^{X+t\sigma} f_{X,\sigma}(x)\, dx.$$

Prove carefully, showing all the necessary changes of variables, that

$$P(\text{within } t\sigma) = \frac{1}{\sqrt{2\pi}} \int_{-t}^{t} e^{-z^2/2}\, dz. \tag{5.69}$$

With each change of variables, check carefully what happens to your limits of integration. The integral (5.69) is often called the error function, denoted erf(t), or normal error integral.

***5.8** (Section 5.4). A student measures some quantity y many times and calculates his mean as $\bar{y} = 23$ and his standard deviation as $\sigma_y = 1$. What fraction of his readings would you expect to find between

(a) 22 and 24?
(b) 22.5 and 23.5?
(c) 21 and 25?
(d) 21 and 23?
(e) 24 and 25?

Finally, (f) within what limits (equidistant on either side of the mean) would you expect to find 50 percent of his readings?

The necessary information for all parts of this question is in Figure 5.13. More detailed information for these kinds of probabilities is in Appendixes A and B.

5.9 (Section 5.4). An extensive survey reveals that the heights of men in a certain country are normally distributed, with mean $\bar{h} = 69''$ and standard deviation $\sigma = 2''$. In a random sample of 1,000 men, how many would you expect to have height

(a) between 67″ and 71″?
(b) more than 71″?
(c) more than 75″?
(d) between 65″ and 67″?

***5.10** (Section 5.5). Suppose we have N measurements $x_1, \ldots, x_N$ of the same quantity x, and we believe that our limiting distribution should be the Gauss function $f_{X,\sigma}(x)$, with X and σ unknown. The principle of maximum likelihood asserts that the best estimate for the width is the value of σ for which the probability $P_{X,\sigma}(x_1, \ldots, x_N)$ of the observed values $x_1, \ldots, x_N$ is largest. Differentiate $P_{X,\sigma}(x_1, \ldots, x_N)$ in (5.41) with respect to σ, and show that the maximum occurs when σ is given by (5.44). As discussed after (5.44), this result means that the best estimate for σ is the standard deviation of the N observed values $x_1, \ldots, x_N$.

5.11 (Section 5.6). Verify the identity (5.54) used in justifying addition in quadrature when propagating random errors.

***5.12** (Section 5.7). Listed here are forty measurements $t_1, \ldots, t_{40}$ of the time for a stone to fall from a window to the ground (all in hundredths of a second).

63	58	74	78	70	74	75	82	68	69
76	62	72	88	65	81	79	77	66	76
86	72	79	77	60	70	65	69	73	77
72	79	65	66	70	74	84	76	80	69

(a) Compute the standard deviation σ_t for the forty measurements.
(b) Compute the means $\bar{t}_1, \ldots, \bar{t}_{10}$ of the four measurements in each of the ten columns. One can now think of the data as resulting from ten experiments, in each of which one found the *mean of four timings*. Given the result of part (a), what would you expect for the standard deviation of the ten averages $\bar{t}_1, \ldots, \bar{t}_{10}$? What is it?
(c) Plot histograms for the forty individual measurements $t_1, \ldots, t_{40}$, and for the ten means $\bar{t}_1, \ldots, \bar{t}_{10}$. Use the same scales and bin sizes for both plots, so that they can be easily compared. Bin boundaries can be chosen in various ways, of which perhaps the simplest is to

put one boundary at the mean of all 40 measurements (72.90), and to use bins whose width is the standard deviation of the ten averages $\bar{t}_1, \ldots, \bar{t}_{10}$.

***5.13** (Section 5.8). A student measures g, the acceleration of gravity, repeatedly and carefully, and gets a final answer of 9.5 m/sec^2 with a standard deviation of 0.1. If his measurement were normally distributed, with center at the accepted value 9.8 and with width 0.1, what would be the probability of getting an answer that differs from 9.8 by as much as (or more than) his? Assuming that he made no actual mistakes, do you think it probable that his experiment suffered from some undetected systematic errors?

5.14 (Section 5.8). Two students measure the same quantity x, and get final answers $x_A = 13 \pm 1$ and $x_B = 15 \pm 1$, where the uncertainties quoted are standard deviations.

(a) Assuming all errors are independent and random, what is the difference $x_A - x_B$ and what is its uncertainty?

(b) Assuming that all quantities were normally distributed as expected, what would be the probability of getting a discrepancy as large as they did? Do you consider their discrepancy significant (at the 5 percent level)?

***5.15** (Section 5.8). An experimenter wishes to check the conservation of energy in a certain nuclear reaction, and measures the initial and final energies as $E_i = 75 \pm 3$ MeV and $E_f = 60 \pm 9$ MeV, where both quoted uncertainties are the standard deviations of the answers. Is this discrepancy significant (5 percent level)? Explain your reasoning clearly.

Part II

If you have read and understood Chapter Five, you are now ready, with surprisingly little difficulty, to study a number of more advanced topics. The seven chapters of Part II present seven such topics, some of them being applications of the statistical theory already developed, others being further extensions of that theory. All of them are important, and the serious student is bound to study them all sooner or later. On the other hand, you might not necessarily want to learn them all at one time. For this reason they have been arranged in independent, short chapters, which can be studied in any order, as your needs and interests dictate.

CHAPTER 6

Rejection of Data

In this chapter we discuss the awkward question of whether to discard a measurement that seems so unreasonable that it looks like a mistake.

6.1. The Problem of Rejecting Data

It sometimes happens that one measurement in a series of measurements appears to disagree strikingly with all the others. When this happens, the experimenter must decide whether the anomalous measurement resulted from some mistake and should be rejected, or was a *bona fide* measurement that should be used with all the others. For example, imagine that we make six measurements of the period of a pendulum and get the results (all in seconds)

$$3.8, 3.5, 3.9, 3.9, 3.4, 1.8. \tag{6.1}$$

In this example the value 1.8 is startlingly different from all the others, and we must decide what to do with it.

We know from Chapter 5 that a legitimate measurement *may* deviate significantly from other measurements of the same quantity. Nevertheless, a legitimate discrepancy as large as that of the last measurement in (6.1) is *very improbable*; so we are inclined to suspect that the time 1.8 sec resulted from some undetected mistake or other external cause. Perhaps we simply misread the last time, or conceivably our electric timer actually stopped briefly during the last measurement because of a momentary power failure.

If we have kept very careful records, we may sometimes be able to establish such a definite cause for the anomalous measurement. For example, our records might show that a different stopwatch was used for the last timing in (6.1), and a subsequent check might show that this watch runs slow. In this case the anomalous measurement should definitely be rejected.

Unfortunately, it is usually not possible to establish an external cause for an anomalous result. We must then decide whether or not to reject the anomaly simply by examining the results themselves, and here our knowledge of the Gauss distribution proves useful.

The rejection of data is a controversial question, on which experts disagree. It is also an *important* question. In our example, the best estimate for the period of the pendulum is significantly affected if we reject the suspect 1.8 sec. The average of all six measurements is 3.4 sec, whereas that of the first five is 3.7 sec, an appreciable difference.

Furthermore, the decision to reject data is ultimately a subjective one, and the scientist who makes this decision may reasonably be accused by other scientists of "fixing" his data. The situation is made worse by the possibility that the anomalous result may reflect some important effect. Indeed, many important scientific discoveries first appeared as anomalous measurements that looked like mistakes. In throwing out the time 1.8 sec in the example (6.1), we just *might* be throwing out the most interesting part of the data.

In fact, faced with data like those in (6.1), our only really honest course is to repeat the measurement many, many times. If the anomaly shows up again, then we will presumably be able to trace its cause, either as a mistake or a real physical effect; if it does not recur, then by the time we have made, say, 100 measurements, there will be no significant difference in our final answer whether we include the anomaly or not.

Nevertheless, it is frequently impractical (especially in a teaching laboratory) to repeat a measurement 100 times every time a result seems suspect. We therefore need some criterion for rejecting a suspect result. There are various such criteria, some quite complicated. The criterion we now describe is called Chauvenet's criterion, and provides a simple and instructive application of the Gauss distribution.

6.2. *Chauvenet's Criterion*

Let us return to the six measurements of the example (6.1):

$$3.8, 3.5, 3.9, 3.9, 3.4, 1.8.$$

If we assume for the moment that these are six legitimate measurements of a quantity x, then we can calculate the mean $\bar{x}$ and standard deviation σ_x,

$$\bar{x} = 3.4 \text{ sec} \tag{6.2}$$

and

$$\sigma_x = 0.8 \text{ sec.} \tag{6.3}$$

We can now quantify the extent to which the suspect measurement, 1.8, is anomalous. It differs from the mean 3.4 by 1.6, or two standard deviations. If we assume that the measurements were governed by a Gauss distribution with center and width given by (6.2) and (6.3), then we can calculate the probability of obtaining measurements that differ by at least this much from the mean. According to the probabilities shown in Figure 5.13, this is

$$\begin{aligned} P(\text{outside } 2\sigma_x) &= 1 - P(\text{within } 2\sigma_x) \\ &= 1 - 0.95 \\ &= 0.05. \end{aligned}$$

In other words, assuming that the values (6.2) and (6.3) for $\bar{x}$ and σ_x are legitimate, we would expect one in every 20 measurements to differ from the mean by at least as much as the suspect 1.8 sec does. If we had made 20 or more measurements, we should actually *expect* to get one or two measurements as bad as the 1.8 sec, and there would be no reason to reject it. But we have made only six measurements; so the expected number of measurements at least as bad as the 1.8 sec was actually

$$0.05 \times 6 = 0.3.$$

That is, in six measurements we would expect (on average) only $\frac{1}{3}$ of a measurement as bad as the suspect 1.8 sec.

This result provides us with the needed quantitative measure of the "reasonableness" of our suspect measurement. If we choose to regard $\frac{1}{3}$ of a measurement as "ridiculously improbable," then we will conclude that the value 1.8 sec was not a legitimate measurement and should be rejected.

The decision where to set the boundary of "ridiculous improbability" is up to the experimenter. Chauvenet's criterion, as normally given, states that if the expected number of measurements at least as bad as the suspect measurement is less than $\frac{1}{2}$, then the suspect measurement should be rejected. Obviously the choice of $\frac{1}{2}$ is arbitrary; but it is also reasonable and can be defended.

The application of Chauvenet's criterion to a general problem can now be easily described. Suppose we make N measurements

$$x_1, \ldots, x_N$$

of the same quantity x. From all N measurements, we calculate $\bar{x}$ and σ_x. If one of the measurements (call it x_{sus}) differs from $\bar{x}$ so much that it looks suspicious, then we first calculate

$$t_{sus} = \frac{x_{sus} - \bar{x}}{\sigma_x}, \tag{6.4}$$

the number of standard deviations by which x_{sus} differs from $\bar{x}$. We next find (from Figure 5.13 or the more complete table in Appendix A) the probability $P(\text{outside } t_{sus}\sigma_x)$ that a legitimate measurement will differ from $\bar{x}$ by t_{sus} or more standard deviations. Finally, we multiply by N, the total number of measurements, to arrive at

$$n(\text{worse than } x_{sus}) = NP(\text{outside } t_{sus}\sigma_x).$$

This n is the number of measurements expected to be at least as bad as x_{sus}. If n is less than $\frac{1}{2}$, then x_{sus} fails Chauvenet's criterion and is rejected.

After rejecting any measurements that fail Chauvenet's criterion, one naturally recalculates $\bar{x}$ and σ_x from the remaining data. The resulting value of σ_x will be smaller than the original one, and it may happen that, with the new σ_x, some more measurements will fail Chauvenet's criterion. However, most authorities agree that Chauvenet's criterion should not be applied a second time using the recalculated values of $\bar{x}$ and σ_x.

Many scientists feel that rejection of data is *never* justified, unless there is *external* evidence that the data in question is incorrect. Perhaps a more moderate position is that Chauvenet's criterion should be used to identify data that at least could be *considered* candidates for rejection. The careful student could then do subsequent calculations twice, once including the questionable data and once without them, to see how much the suspect values actually affect the final conclusion.

6.3. An Example

A student makes ten measurements of one length x and gets the results (all in mm)

$$46, 48, 44, 38, 45, 47, 58, 44, 45, 43.$$

Noticing that the value 58 seems anomalously large, he checks his records, but can find no evidence that the result was caused by a mistake. He therefore applies Chauvenet's criterion. What does he conclude?

Accepting provisionally all ten measurements, he computes

$$\bar{x} = 45.8 \quad \text{and} \quad \sigma_x = 5.1.$$

The difference between the suspect value $x_{\text{sus}} = 58$ and the mean $\bar{x} = 45.8$ is 12.2, or 2.4 standard deviations; i.e.,

$$\frac{x_{\text{sus}} - \bar{x}}{\sigma_x} = \frac{58 - 45.8}{5.1} = 2.4.$$

Referring to the table in Appendix *A*, he sees that the probability that a measurement will differ from $\bar{x}$ by 2.4 σ_x or more is

$$\begin{aligned} P(\text{outside } 2.4\sigma) &= 1 - P(\text{within } 2.4\sigma) \\ &= 1 - 0.984 \\ &= 0.016. \end{aligned}$$

In ten measurements he would therefore expect to find only .16 of one measurement as bad as his suspect result. Since this is less than the number $\frac{1}{2}$ set by Chauvenet's criterion, he should at least consider rejecting the result.

His next most suspect result is 38, which is 1.5 standard deviations away from the mean $\bar{x} = 45.8$. A similar calculation shows that, among ten measurements, he would expect 1.3 as bad as this; so this result is entirely acceptable. If he decides to reject the suspect 58, then he must recalculate $\bar{x}$ and σ_x as

$$\bar{x} = 44.4 \quad \text{and} \quad \sigma_x = 2.9.$$

As one would expect, his mean changes a bit, and his standard deviation drops appreciably.

Problems

Reminder: An asterisk indicates that the problem is discussed, or its answer given, in the Answers section at the back of the book.

6.1 (Section 6.2). An enthusiastic student makes 50 measurements of the heat Q released in a certain process. Her average and standard deviation are $\bar{Q} = 4.8$ and $\sigma_Q = 0.4$, both in calories.

(a) Assuming that her measurements are governed by the normal distribution, find the probability that any one measurement would differ from $\bar{Q}$ by 0.8 cal or more. How many measurements should she expect to differ from $\bar{Q}$ by 0.8? If one of her measurements is 4.0 cal, and she decides to use Chauvenet's criterion, would she reject this measurement?

(b) If one of her measurements is 6.0 cal, would she reject it?

***6.2** (Section 6.2). A student measures a certain voltage V ten times, with the results (in volts)

$$.86, .83, .87, .84, .82, .95, .83, .85, .89, .88.$$

(a) Calculate the mean $\bar{V}$ and standard deviation σ_V of these results.

(b) If he decides to use Chauvenet's criterion, should he reject the reading of .95 volts? Explain your reasoning clearly.

***6.3** (Section 6.2). A student makes 14 measurements of the period of a damped oscillator, with the results (in tenths of a second)

$$7, 3, 9, 3, 6, 9, 8, 7, 8, 12, 5, 9, 9, 3.$$

Feeling that the result 12 is suspiciously high, she decides to apply Chauvenet's criterion. Does she reject the suspect result? How many results should she expect to be as far from the mean as 12 is?

6.4 (Section 6.2). Chauvenet's criterion defines a boundary outside which a measurement is regarded as rejectable. If we make ten measurements and one of them differs from the mean by more than about two standard deviations (in either direction), then that measurement is considered as rejectable; for 20 measurements, the corresponding boundary is about 2.2 standard deviations. Make a table showing the "boundary of rejectability" for 5, 10, 15, 20, 50, 100, 200, and 1,000 measurements. (Use the table of the error function in Appendix A.)

CHAPTER 7

Weighted Averages

In this chapter we address the problem of combining two or more separate and independent measurements of a single physical quantity. We will find that the best estimate of that quantity, based on the several measurements, is an appropriate *weighted average* of those measurements.

7.1. The Problem of Combining Separate Measurements

It often happens that a physical quantity is measured several times, perhaps in several separate laboratories, and the question arises how these measurements can be combined to give a single best estimate. Suppose, for example, that two students, A and B, measure a quantity x carefully and obtain these results:

$$\text{Student } A; \qquad x = x_A \pm \sigma_A \tag{7.1}$$

and

$$\text{Student } B; \qquad x = x_B \pm \sigma_B. \tag{7.2}$$

Each of these results will probably itself be the result of several measurements, in which case x_A will be the mean of all A's measurements and σ_A the standard deviation of that mean (and similarly for x_B and σ_B). The question is how best to combine x_A and x_B for a single best estimate of x.

Before we answer this question, let us notice that if the discrepancy $|x_A - x_B|$ between the two measurements is much greater than both uncertainties σ_A and σ_B, then we should suspect that something has gone wrong in at least one of the measurements. In this situation we would say that the two measurements are *inconsistent*, and we should examine both measurements carefully to see whether either (or both) was subject to unnoticed systematic errors.

Let us suppose, however, that the two measurements (7.1) and (7.2) are *consistent*; that is, the discrepancy $|x_A - x_B|$ is *not* significantly larger than both σ_A and σ_B. It then makes sense to ask what is the best estimate x_{best} of the true value X, based on the two measurements. One's first impulse might be to use the average $(x_A + x_B)/2$ of the two measurements. However, a little reflection should suggest that this is unsuitable if the two uncertainties σ_A and σ_B are unequal. The simple average $(x_A + x_B)/2$ gives equal importance to both measurements, whereas the reading that is more precise should somehow be given more weight.

7.2. The Weighted Average

We can solve our problem easily by using the principle of maximum likelihood, much as we did in Section 5.5. If we assume that both measurements are governed by the Gauss distribution and we denote the unknown true value of x by X, then the probability of student A's obtaining his particular value x_A is

$$P_X(x_A) \propto \frac{1}{\sigma_A} e^{-(x_A - X)^2/2\sigma_A{}^2}, \tag{7.3}$$

and that of B's getting his observed x_B is

$$P_X(x_B) \propto \frac{1}{\sigma_B} e^{-(x_B - X)^2/2\sigma_B{}^2}. \tag{7.4}$$

We have indicated explicitly, with the subscript X, that these probabilities depend on the unknown actual value. (They also depend on the respective widths, σ_A and σ_B, but we haven't indicated this.)

The probability that A find the value x_A *and* B the value x_B is just the product of the two probabilities (7.3) and (7.4). In a way that should now be familiar, this product will involve an exponential function whose exponent is the sum of the two exponents in (7.3) and (7.4). We write this as

$$\begin{aligned} P_X(x_A, x_B) &= P_X(x_A)P_X(x_B) \\ &\propto \frac{1}{\sigma_A \sigma_B} e^{-\chi^2/2}, \end{aligned} \tag{7.5}$$

where we have introduced the convenient shorthand χ^2 (chi squared) for the exponent

$$\chi^2 = \left(\frac{x_A - X}{\sigma_A}\right)^2 + \left(\frac{x_B - X}{\sigma_B}\right)^2. \tag{7.6}$$

This important quantity is the sum of the squares of the deviations from X of the two measurements, each divided by its corresponding uncertainty. It is sometimes called just the "sum of squares."

The principle of maximum likelihood asserts, just as before, that our best estimate for the unknown true value X is that value for which the actual observations x_A, x_B are most likely. That is, the best estimate for X is the value for which the probability (7.5) is maximum or, equivalently, the exponent χ^2 is minimum. (Since maximizing the probability entails minimizing the "sum of squares" χ^2, this method for estimating X is sometimes called the "method of least squares.") Thus, to find the best estimate, we simply differentiate (7.6) with respect to X and set the derivative equal to zero,

$$2\frac{x_A - X}{{\sigma_A}^2} + 2\frac{x_B - X}{{\sigma_B}^2} = 0.$$

The solution of this equation for X is the best estimate x_{best} and is easily seen to be

$$x_{\text{best}} = \left(\frac{x_A}{{\sigma_A}^2} + \frac{x_B}{{\sigma_B}^2}\right) \Big/ \left(\frac{1}{{\sigma_A}^2} + \frac{1}{{\sigma_B}^2}\right). \tag{7.7}$$

This rather ugly result can be made tidier if we define *weights*

$$w_A = \frac{1}{{\sigma_A}^2} \quad \text{and} \quad w_B = \frac{1}{{\sigma_B}^2}. \tag{7.8}$$

Substituting in (7.7), we obtain

$$x_{\text{best}} = \frac{w_A x_A + w_B x_B}{w_A + w_B}. \tag{7.9}$$

If the original two measurements are equally precise ($\sigma_A = \sigma_B$ and hence $w_A = w_B$), our answer reduces to the simple average $(x_A + x_B)/2$. In general, (7.9) is a *weighted average*; it is similar to the formula for the center of gravity of two bodies, where w_A and w_B are the actual weights of the two bodies, and x_A and x_B their positions. Here the "weights" are the inverse squares of the uncertainties in the original measurements, as in (7.8). If A's measurement is more precise than B's, then $\sigma_A < \sigma_B$ and hence $w_A > w_B$; so the best estimate x_{best} is closer to x_A than to x_B, just as it should be.

Our analysis can be generalized to combine several measurements of a single quantity. Suppose we have N separate measurements of a quantity x,

$$x_1 \pm \sigma_1, \quad x_2 \pm \sigma_2, \ldots, \quad x_N \pm \sigma_N$$

with their corresponding uncertainties $\sigma_1, \ldots, \sigma_N$. Arguing much as before, we find that the best estimate based on these measurements is the weighted average

$$x_{\text{best}} = \frac{\sum_{i=1}^{N} w_i x_i}{\sum_{i=1}^{N} w_i}, \tag{7.10}$$

where the *weights* w_i are the inverse squares of the corresponding uncertainties,

$$w_i = 1/\sigma_i^2, \tag{7.11}$$

for $i = 1, 2, \ldots, N$.

Since the weight $w_i = 1/\sigma_i^2$ attached to each measurement involves the *square* of the corresponding uncertainty σ_i, any measurement which is much less precise than the others contributes very much less to the final answer (7.10). For example, if one measurement is four times less precise than the rest, then its weight is 16 times less than the other weights, and for many purposes this measurement could simply be ignored.

Since the final answer (7.10) for x_{best} is a simple function of the original measured values $x_1, \ldots, x_N$, it is a simple matter to calculate the uncertainty in our answer using error propagation. Problem 7.5 will ask you to check that the uncertainty in the answer (7.10) for x_{best} is

$$\sigma_{x_{\text{best}}} = \left(\sum_{i=1}^{N} w_i \right)^{-1/2}, \tag{7.12}$$

where $w_i = 1/\sigma_i^2$, as usual.

7.3. An Example

Three students measure a resistance several times and obtain the following three answers (in ohms):

(first student's value for R) $= 11 \pm 1$;
(second student's value for R) $= 12 \pm 1$;
(third student's value for R) $= 10 \pm 3$.

Given these three results, what is the best estimate for the resistance R?

The three uncertainties σ_1, σ_2, σ_3 are 1, 1, and 3. The three weights, $w_i = 1/\sigma_i^2$, are therefore

$$w_1 = 1, \qquad w_2 = 1, \qquad w_3 = \tfrac{1}{9}.$$

Thus according to (7.10) the best estimate is

$$R_{\text{best}} = \frac{\sum w_i R_i}{\sum w_i} = \frac{(1 \times 11) + (1 \times 12) + (\frac{1}{9} \times 10)}{1 + 1 + \frac{1}{9}}$$

$$= 11.42 \text{ ohms}.$$

The uncertainty in this answer is given by (7.12) as

$$\sigma_{R_{\text{best}}} = \left(\sum w_i\right)^{-1/2} = \left(1 + 1 + \tfrac{1}{9}\right)^{-1/2} = 0.69.$$

Thus our final conclusion is

$$R = 11.4 \pm 0.7 \text{ ohms}.$$

It is interesting to see what answer we would get if we were to ignore completely the third student's measurement, which is three times less accurate and hence nine times less important. Here a simple calculation gives $R_{\text{best}} = 11.50$ (compared with 11.42) with an uncertainty of 0.71 (compared with 0.69). Obviously the third measurement does not have a big effect.

Problems

Reminder: An asterisk indicates that the problem is discussed, or its answer given, in the Answers section at the back of the book.

***7.1** (Section 7.2).

(a) Two measurements of the speed of sound u give the answers 334 ± 1

and 336 ± 2 (both in m/sec). Would you consider them consistent? If so, calculate the best estimate for u and its uncertainty.

(b) Repeat part (a) for the results 334 ± 1 and 336 ± 5. Is the second result here worth including?

***7.2** (Section 7.2). Two students measure a resistance by different methods. Each makes ten measurements and computes the mean and its standard deviation, with these results:

$$\text{Student } A, \quad R = 72 \pm 8 \text{ ohms};$$
$$\text{Student } B, \quad R = 78 \pm 5 \text{ ohms}.$$

(a) Including both measurements, what are the best estimate of R and its uncertainty?

(b) About how many measurements would student A have to make (using his same technique) to give his result the same weight as that of B?

7.3 (Section 7.2). Find the best estimate and its uncertainty based on the following four measurements of one quantity:

$$1.4 \pm .5, \qquad 1.2 \pm .2, \qquad 1.0 \pm .25, \qquad 1.3 \pm .2.$$

7.4 (Section 7.2). Suppose that N measurements of some quantity x all have the same uncertainty. Show clearly that in this situation the weighted average (7.10) reduces to the ordinary average, or mean, $\bar{x} = (\sum x_i)/N$, and that the expression (7.12) for the uncertainty reduces to the familiar standard deviation of the mean.

***7.5** (Section 7.2). Given N measurements $x_1, \ldots, x_N$ of one quantity x with uncertainties $\sigma_1, \ldots, \sigma_N$, the best estimate for x is given by (7.10) as $x_{\text{best}} = (\sum w_i x_i)/(\sum w_i)$, with weights $w_i = 1/\sigma_i^2$. This defines x_{best} as a function of $x_1, \ldots, x_N$. Use the formula (3.47) for error propagation to show that the uncertainty in x_{best} is given by (7.12) as

$$\sigma_{x_{\text{best}}} = (\textstyle\sum w_i)^{-1/2}.$$

CHAPTER 8

Least-Squares Fitting

In our discussion of the statistical analysis of data, we have so far focused exclusively on the repeated measurement of one single quantity, not because the analysis of many measurements of one quantity is the most interesting problem in statistics, but because this simple problem must be well-understood before we can discuss more general ones. We are at last ready to discuss our first, and very important, more general problem.

8.1. Data that Should Fit a Straight Line

One of the most common and interesting types of experiment involves the measurement of several values of two different physical variables, in order to investigate the mathematical relationship between the two variables. For instance, one might drop a stone from various different heights $h_1, \ldots, h_N$ and measure the corresponding times of fall $t_1, \ldots, t_N$ to see if the heights and times are connected by the expected relation $h = \frac{1}{2}gt^2$.

Probably the most important experiments of this type are those where the expected relation is *linear*, and this is the case we consider first. For instance, if we believe that a body is falling with constant acceleration g, then its velocity v should be a linear function of the time t,

$$v = v_0 + gt.$$

More generally, we will consider any two physical variables x and y that we suspect are connected by a linear relation of the form

$$y = A + Bx, \tag{8.1}$$

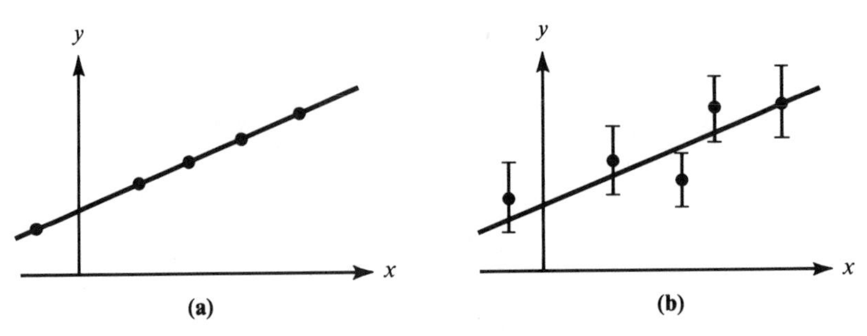

Figure 8.1. (a) If the two variables y and x are linearly related as in Equation (8.1), and if there were no experimental uncertainties, then the measured points (x_i, y_i) would all lie exactly on the line $y = A + Bx$. (b) In practice there always are uncertainties, which can be shown by error bars, and the points (x_i, y_i) can be expected only to lie reasonably close to the line. Here only y is shown as subject to appreciable uncertainties.

where A and B are constants. Unfortunately, many different notations are used for a linear relation; beware of confusing the form (8.1) with the equally popular $y = ax + b$.

If the two variables y and x are linearly related as in (8.1), then a graph of y against x should be a straight line which has slope B and intersects the y axis at $y = A$. If we were to measure N different values $x_1, \ldots, x_N$ and the corresponding values $y_1, \ldots, y_N$, and if our measurements were subject to no uncertainties, then each of the points (x_i, y_i) would lie exactly on the line $y = A + Bx$, as in Figure 8.1(a). In practice, there *are* uncertainties, and the most we can expect is that the distance of each point (x_i, y_i) from the line will be reasonable compared to the uncertainties, as in Figure 8.1(b).

When we make a series of measurements of the kind just described, there are two possible questions we can ask. First, if we take for granted that y and x *are* linearly related, then the interesting problem is to find the straight line $y = A + Bx$ that best fits the measurements; that is, to find the best estimates for the constants A and B based on the data $(x_1, y_1), \ldots, (x_N, y_N)$. This problem can be approached graphically, as discussed briefly in Section 2.6. It can also be treated analytically, by means of the principle of maximum likelihood. This analytical method of finding the best straight line to fit a series of experimental points is called *linear regression*, or the *least-squares fit for a line*, and is the main subject of this chapter.

The second question that can be asked is whether the measured values $(x_1, y_1), \ldots, (x_N, y_N)$ do really bear out our expectation that y is linear in

x. We can first find the line that best fits the data, but we must then devise some measure of *how well* this line fits the data. We will take up this second question in Chapter 9.

8.2. Calculation of the Constants A and B

Let us now return to the question of finding the best straight line $y = A + Bx$ to fit a set of measured points $(x_1, y_1), \ldots, (x_N, y_N)$. To simplify our discussion, we will suppose that, although our measurements of y suffer some uncertainty, the uncertainty in our measurements of x is negligible. This is often a reasonable assumption, since the uncertainties in one variable often are much larger than those in the other, which we can therefore safely ignore. We will further assume that the uncertainties in y all have the same magnitude. (This is also a reasonable assumption in many experiments, but if the uncertainties are different, then our analysis can be generalized to weight the measurements appropriately; see Problem 8.4.) More specifically, we assume that the measurement of each y_i is governed by the Gauss distribution, with the same width parameter σ_y for all measurements.

If we knew the constants A and B, then, for any given value x_i (which we are assuming has no uncertainty), we could compute the true value of the corresponding y_i,

$$(\text{true value for } y_i) = A + Bx_i. \tag{8.2}$$

The measurement of y_i is governed by a normal distribution centered on this true value, with width parameter σ_y. Therefore, the probability of obtaining the observed value y_i is

$$P_{A,B}(y_i) \propto \frac{1}{\sigma_y} e^{-(y_i - A - Bx_i)^2/2\sigma_y{}^2}, \tag{8.3}$$

where the subscripts A and B indicate that this probability depends on the (unknown) values of A and B. The probability of obtaining our complete set of measurements $y_1, \ldots, y_N$ is the product

$$\begin{aligned} P_{A,B}(y_1, \ldots, y_N) &= P_{A,B}(y_1) \cdots P_{A,B}(y_N) \\ &\propto \frac{1}{\sigma_y{}^N} e^{-\chi^2/2}, \end{aligned} \tag{8.4}$$

where the exponent is given by

$$\chi^2 = \sum_{i=1}^{N} \frac{(y_i - A - Bx_i)^2}{\sigma_y^{\,2}}. \tag{8.5}$$

In the now familiar way, the best estimates for the unknown constants A and B, based on the given measurements, are those values of A and B for which the probability $P_{A,B}(y_1, \ldots, y_N)$ is maximum, or for which the sum of squares χ^2 in (8.5) is a minimum (which is why the method is known as least-squares fitting). To find these values, we differentiate χ^2 with respect to A and B and set the derivatives equal to zero:

$$\frac{\partial \chi^2}{\partial A} = (-2/\sigma_y^{\,2}) \sum_{i=1}^{N} (y_i - A - Bx_i) = 0 \tag{8.6}$$

and

$$\frac{\partial \chi^2}{\partial B} = (-2/\sigma_y^{\,2}) \sum_{i=1}^{N} x_i(y_i - A - Bx_i) = 0. \tag{8.7}$$

These two equations can be rewritten as simultaneous equations for A and B:

$$AN + B\sum x_i = \sum y_i \tag{8.8}$$

and

$$A\sum x_i + B\sum x_i^2 = \sum x_i y_i. \tag{8.9}$$

(From now on we omit the limits $i = 1$ to N from the summation signs $\sum$.) These two equations, known as *normal equations*, are easily solved to give the *least-squares estimates for the constants A and B*

$$A = \frac{(\sum x_i^2)(\sum y_i) - (\sum x_i)(\sum x_i y_i)}{\Delta} \tag{8.10}$$

and

$$B = \frac{N(\sum x_i y_i) - (\sum x_i)(\sum y_i)}{\Delta}, \tag{8.11}$$

where we have introduced the convenient abbreviation

$$\Delta = N(\sum x_i^2) - (\sum x_i)^2. \tag{8.12}$$

The results (8.10) and (8.11) give the best estimates for the constants A and B of the straight line $y = A + Bx$, based on the measured points $(x_1, y_1), \ldots, (x_N, y_N)$. The resulting line is called the *least-squares fit* to the data, or the *line of regression* of y on x. It is now natural to ask what are the uncertainties in our estimates for A and B. It turns out that before we can answer this question, we must discuss the uncertainty σ_y in our original measurements of $y_1, \ldots, y_N$, and this we take up next.

8.3. Uncertainty in the Measurements of y

In the course of measuring the values $y_1, \ldots, y_N$, we have presumably formed some idea of their uncertainty. Nonetheless, it is important to know how to calculate the uncertainty by analyzing the data themselves. One must remember that the numbers $y_1, \ldots, y_N$ are *not* N measurements of the same quantity. (They might, for instance, be the times for a stone to fall from N different heights.) Thus we certainly do not get an idea of their reliability by examining the spread in their values.

Nevertheless, we can easily estimate the uncertainty σ_y in the numbers $y_1, \ldots, y_N$. The measurement of each y_i is (we are assuming) normally distributed about its true value $A + Bx_i$, with width parameter σ_y. Thus the *deviations* $y_i - A - Bx_i$ are normally distributed, all with the same central value 0 and the same width σ_y. This immediately suggests that a good estimate for σ_y would be given by a sum of squares with the familiar form

$$\sigma_y^2 = \frac{1}{N} \sum (y_i - A - Bx_i)^2. \tag{8.13}$$

In fact, this answer can be confirmed by means of the principle of maximum likelihood. As usual, the best estimate for the parameter in question (σ_y here) is that value for which the probability (8.4) of obtaining the observed values $y_1, \ldots, y_N$ is maximum. As you can easily check by differentiating (8.4) with respect to σ_y and setting the derivative equal to zero, this best estimate is precisely the answer (8.13).

Unfortunately, as you may have suspected, the estimate (8.13) for $\sigma_y{}^2$ is not quite the end of the story. The numbers A and B in (8.13) are the unknown true values of the constants A and B. In practice, these must be replaced by our *best estimates* for A and B, namely, (8.10) and (8.11), and this replacement slightly reduces the value of (8.13). It can be shown that this reduction is compensated for if we replace the factor N in the denominator by $(N - 2)$. Thus our final answer for the uncertainty in the measurements $y_1, \ldots, y_N$ is

$$\sigma_y{}^2 = \frac{1}{N-2} \sum_{i=1}^{N} (y_i - A - Bx_i)^2, \tag{8.14}$$

with A and B given by (8.10) and (8.11). If we already have an independent estimate of our uncertainty in $y_1, \ldots, y_N$, then we would expect this estimate to compare with σ_y as computed from (8.14).

We will not attempt to justify the factor of $(N - 2)$ in (8.14), but we can make some comments. First, as long as N is moderately large the difference between N and $(N - 2)$ is unimportant anyway. Second, that the factor $(N - 2)$ is *reasonable* becomes clear if we consider measuring just two pairs of data (x_1, y_1) and (x_2, y_2). With only two points, we can always find a line that passes *exactly* through both points, and the least-squares fit will give this line. That is, with just two pairs of data, we cannot possibly deduce anything about the reliability of our measurements. Now, since both points lie exactly on the best line, the two terms of the sum in (8.13) and (8.14) are zero. Thus the formula (8.13) (with $N = 2$ in the denominator) would give the absurd answer $\sigma_y = 0$; whereas (8.14), with $N - 2 = 0$ in the denominator, gives $\sigma_y = 0/0$, indicating correctly that σ_y is undetermined after only two measurements.

The presence of the factor $(N - 2)$ in (8.14) is reminiscent of the $(N - 1)$ that appeared in our estimate of the standard deviation of N measurements of one quantity x, in Equation (5.46). There we made N measurements $x_1, \ldots, x_N$ of the one quantity x. Before we could calculate σ_x, we had to use our data to find the mean $\bar{x}$. In a certain sense, this left only $(N - 1)$ independent measured values; so we say that, having computed $\bar{x}$, we have only $(N - 1)$ *degrees of freedom* left. Here we made N measurements, but before calculating σ_y we had to compute the *two* quantities A and B. Having done this, we had only $(N - 2)$ degrees of freedom left. In general, we define the *number of degrees of freedom* at any stage in a statistical calculation as the number of independent measurements *minus* the number of parameters calculated from these measurements. It is possible to show (though we will not do so here) that it is the number of degrees of freedom, *not* the number of measurements, that should appear in the denominator of formulas like (8.14) and (5.46). This explains why (8.14) contains the factor $(N - 2)$, and (5.46) the factor $(N - 1)$.

8.4. Uncertainty in the Constants A and B

Having found the uncertainty σ_y in the measured numbers $y_1, \ldots, y_N$, we can easily return to our estimates for the constants A and B and calculate their uncertainties. The point is that the estimates (8.10) and (8.11) for A and B are well-defined functions of the measured numbers $y_1, \ldots, y_N$. Therefore the uncertainties in A and B are given by simple error propagation in terms of those in $y_1, \ldots, y_N$. We leave it to the reader to check (Problem 8.8) that

$$\sigma_A{}^2 = \sigma_y{}^2 \sum x_i{}^2/\Delta \tag{8.15}$$

and

$$\sigma_B{}^2 = N\sigma_y{}^2/\Delta, \tag{8.16}$$

where Δ is given by (8.12) as usual.

8.5. An Example

If the volume of a sample of an ideal gas is kept constant, then its temperature T is a linear function of its pressure P,

$$T = A + BP. \tag{8.17}$$

Here the constant A is the temperature at which the pressure P would drop to zero (if the gas did not condense into a liquid first); it is called the *absolute zero of temperature*, and has the accepted value

$$A = -273.15 \text{ degrees Celsius.} \tag{8.18}$$

The constant B depends on the nature of the gas, its mass, and its volume.[1] By measuring a series of values for T and P, we can find the best estimates for the constants A and B. In particular, the value of A gives the absolute zero of temperature.

One set of five measurements of P and T obtained by a student was as shown in the first three columns of Table 8.1. The student judged that his measurements of P had negligible uncertainty, and those of T were all equally uncertain with an uncertainty of "a few degrees." Assuming that his points should fit a straight line of the form (8.17), he calculated his best estimate for the constant A (the absolute zero) and its uncertainty. What should have been his conclusions?

Table 8.1. Pressure-temperature experiment.

Trial number, i	Pressure, P_i (in mm of mercury)	Temperature, T_i (in °C)	$A + BP_i$
1	65	−20	−22.2
2	75	17	14.9
3	85	42	52.0
4	95	94	89.1
5	105	127	126.2

All we have to do here is use formulas (8.10) and (8.15), with x_i replaced by P_i and y_i by T_i, to calculate all the quantities of interest. This requires us to compute the sums $\sum P_i$, $\sum P_i^2$, $\sum T_i$, $\sum P_i T_i$. Many pocket calculators can evaluate all these sums automatically; but even without such a machine, we can easily handle these calculations if the data are properly organized. From Table 8.1 we can calculate

$$\begin{aligned} \textstyle\sum P_i &= 425, \\ \textstyle\sum P_i^2 &= 37{,}125, \\ \textstyle\sum T_i &= 260, \\ \textstyle\sum P_i T_i &= 25{,}810, \\ \Delta &= 5{,}000, \end{aligned}$$

where $\Delta = N(\sum P_i^2) - (\sum P_i)^2$. In this kind of calculation, it is important to keep plenty of significant figures, since we have to take differences of these large numbers. Armed with these sums, we can immediately calculate

[1] The difference $T - A$ is called the *absolute temperature*. Thus (8.17) can be rewritten to say that the absolute temperature is proportional to the pressure (at constant volume).

the best estimates for the constants A and B:

$$A = \frac{(\sum P_i^2)(\sum T_i) - (\sum P_i)(\sum P_i T_i)}{\Delta} = -263.35$$

and

$$B = \frac{N(\sum P_i T_i) - (\sum P_i)(\sum T_i)}{\Delta} = 3.71.$$

This already gives the student's best estimate for absolute zero, $A = -263$ degrees Celsius.

Knowing the constants A and B, we can next calculate the numbers $A + BP_i$, the temperatures "expected" on the basis of our best fit to the relation $T = A + BP$. These are shown in the right-hand column of the table, and as we would hope, all agree reasonably well with the observed temperatures. We can now take the difference between the figures in the last two columns and calculate

$$\sigma_T^2 = \frac{1}{N-2}\sum(T_i - A - BP_i)^2 = 44.6$$

and hence the standard deviation,

$$\sigma_T = 6.7.$$

This agrees reasonably with the student's estimate that his temperature measurements were uncertain by "a few degrees."

Finally, we can calculate the uncertainty in A using (8.15):

$$\sigma_A^2 = \sigma_T^2(\sum P_i^2)/\Delta = 331$$

or

$$\sigma_A = 18.$$

Thus our student's final conclusion, suitably rounded, should be

absolute zero, $A = -260 \pm 20$ degrees Celsius,

which agrees satisfactorily with the accepted value, -273 degrees.

As is often true, these results become much clearer if we graph them, as in Figure 8.2. The five data points, with their uncertainties of $\pm 7°$ in T, are shown on the upper right. The best straight line passes through four of the error bars and close to the fifth.

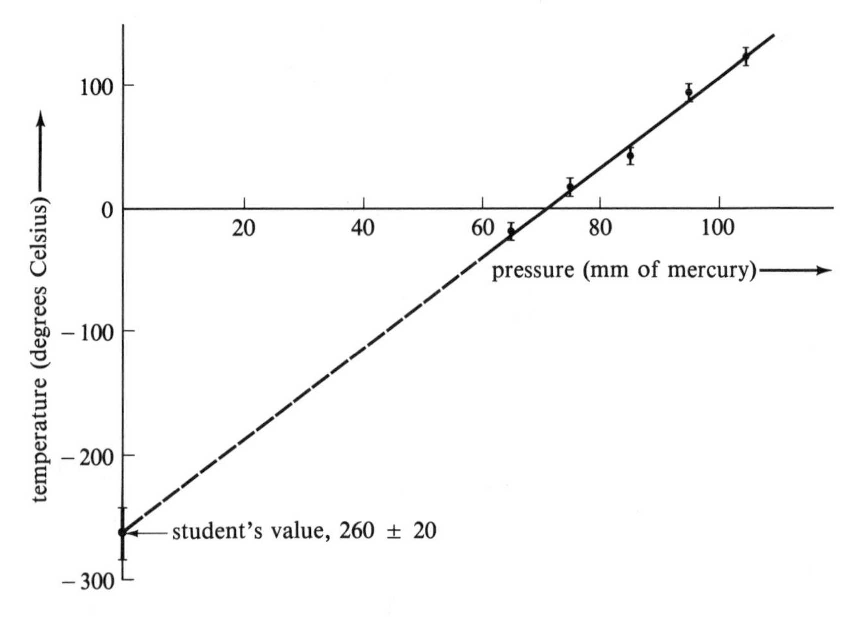

Figure 8.2. Graph of temperature T vs. pressure P for a gas at constant volume. The error bars extend one standard deviation, σ_T, on each side of the five experimental points, and the line is the least-squares best fit. The absolute zero of temperature has been found by extrapolating the line back to its intersection with the T axis.

In order to find a value for absolute zero, the line has been extended beyond all the data points, to its intersection with the T axis. This process of *extrapolation* (extending a curve beyond the data points that determine it) can introduce large uncertainties, as is clear from the picture. A very small change in the line's slope will cause a large change in its intercept on the distant T axis. Thus any uncertainty in the data is greatly magnified if we have to extrapolate any distance. This explains why the uncertainty in the value of absolute zero ($\pm 18°$) is so much larger than that in the original temperature measurements ($\pm 7°$).

8.6. *Least-Squares Fits to Other Curves*

So far in this chapter we have considered the observation of two variables satisfying a linear relation, $y = A + Bx$, and we have discussed the calculation of the constants A and B. This important problem is a special case of a wide class of curve-fitting problems, many of which can be solved in a similar way. In this last section we mention briefly a few more of these problems.

Fitting a Polynomial

It often happens that one variable, y, is expected to be expressible as a polynomial in a second variable x,

$$y = A + Bx + Cx^2 + \cdots + Hx^n. \tag{8.19}$$

For example, the height y of a falling body is expected to be quadratic in the time t,

$$y = y_0 + v_0 t - \tfrac{1}{2}gt^2,$$

where y_0 and v_0 are the initial height and velocity, and g is the acceleration of gravity. Given a set of observations of the two variables, one can find best estimates for the constants $A, B, \ldots, H$ in (8.19) by an argument which exactly parallels that of Section 8.2, as we now outline.

To simplify matters, we suppose that the polynomial (8.19) is actually a quadratic,

$$y = A + Bx + Cx^2. \tag{8.20}$$

(The interested reader can easily extend the analysis to the general case.) We suppose, as before, that we have a series of measurements (x_i, y_i), $i = 1, \ldots, N$, with the y_i all equally uncertain and the x_i all exact. For each x_i, the corresponding true value of y_i is given by (8.20), with A, B, and C as yet unknown. We assume that the measurements of the y_i are governed by normal distributions, each centered on the appropriate true value and all with the same width σ_y. This lets us compute the probability of obtaining our observed values $y_1, \ldots, y_N$ in the familiar form

$$P(y_1, \ldots, y_N) \propto e^{-\chi^2/2}, \tag{8.21}$$

where now

$$\chi^2 = \sum_{i=1}^{N} \frac{(y_i - A - Bx_i - Cx_i{}^2)^2}{\sigma_y{}^2}. \tag{8.22}$$

(This corresponds to Equation (8.5) for the linear case.) The best estimates for A, B, and C are those values for which $P(y_1, \ldots, y_N)$ is largest, or χ^2 is smallest. Differentiating χ^2 with respect to A, B, and C and setting these derivatives equal to zero, we obtain the three equations (as you should check):

$$\begin{aligned} AN + B\sum x_i + C\sum x_i^2 &= \sum y_i, \\ A\sum x_i + B\sum x_i^2 + C\sum x_i^3 &= \sum x_i y_i, \\ A\sum x_i^2 + B\sum x_i^3 + C\sum x_i^4 &= \sum x_i^2 y_i. \end{aligned} \tag{8.23}$$

For any given set of measurements (x_i, y_i), these simultaneous equations for A, B, and C (known as the *normal equations*) can be solved to find the best estimates for A, B, and C. With A, B, C calculated in this way, the equation $y = A + Bx + Cx^2$ is called the least-squares polynomial fit, or the polynomial regression, for the given measurements.

The method of polynomial regression generalizes easily to a polynomial of any degree, although the resulting normal equations become very cumbersome for polynomials of high degree. In principle, a similar method can be applied to *any* function $y = f(x)$ that depends on various unknown parameters $A, B, \ldots$. Unfortunately, the resulting normal equations that determine the best estimates for $A, B, \ldots$ can be difficult or impossible to solve. However, there is one large class of problems which *can* always be solved, namely, those problems where the function $y = f(x)$ depends linearly on the parameters $A, B, \ldots$. These include all polynomials—obviously the polynomial (8.19) is linear in its coefficients $A, B, \ldots$—but they include many other functions. For example, in some problems y is expected to be a sum of trigonometric functions, like

$$y = A \sin x + B \cos x. \tag{8.24}$$

For this function, and in fact for any function that is linear in the parameters $A, B, \ldots$, the normal equations that determine the best estimates for $A, B, \ldots$ are simultaneous linear equations, which can always be solved (see Problems 8.12 and 8.13).

Exponential Functions

One of the most important functions in physics is the exponential function

$$y = Ae^{Bx}, \tag{8.25}$$

where A and B are constants. The intensity I of radiation, after passing a distance x through a shield, falls off exponentially:

$$I = I_0 e^{-\mu x},$$

where I_0 is the original intensity and μ characterizes the absorption by the shield. The charge on a short-circuited capacitor drains away exponentially:

$$Q = Q_0 e^{-\lambda t}$$

where Q_0 is the original charge and $\lambda = 1/(RC)$, R and C being the resistance and capacitance.

If the constants A and B in (8.25) are unknown, then it is natural to seek estimates of them based on measurements of x and y. Unfortunately, direct application of our previous arguments leads to equations for A and B that cannot be conveniently solved. However, it is possible to transform the nonlinear relation (8.25) between y and x into a linear relation, to which we can apply our least-squares fit.

To effect the desired "linearization," we simply take the logarithm of (8.25) to give

$$\ln y = \ln A + Bx. \tag{8.26}$$

We see that, even though y is not linear in x, $\ln y$ *is*. This conversion of the nonlinear (8.25) into the linear (8.26) is useful in many contexts besides that of least-squares fitting. If we wish to check the relation (8.25) graphically, then a direct plot of y against x will produce a curve that is hard to identify visually. On the other hand, a plot of $\ln y$ against x (or of $\log y$ against x) should produce a straight line, which can be easily identified. (Such a plot is especially easy if one uses "semilog" graph paper, on which the graduations on one axis are spaced logarithmically. Such paper lets one plot $\log y$ directly without even calculating it.)

The usefulness of the linear equation (8.26) in least-squares fitting is readily apparent. If we believe that y and x should satisfy $y = Ae^{Bx}$, then the variables $z = \ln y$ and x should satisfy (8.26), or

$$z = \ln A + Bx. \tag{8.27}$$

If we have a series of measurements (x_i, y_i), then for each y_i we can calculate $z_i = \ln y_i$. Then the pairs (x_i, z_i) should lie on the line (8.27). This line can be fitted by the method of least squares to give best estimates for the constants $\ln A$ (from which we can find A) and B.

Example

Many populations (of people, of bacteria, of radioactive nuclei, etc.) tend to vary exponentially in time. If a population N is decreasing exponentially, we write

$$N = N_0 e^{-t/\tau}, \tag{8.28}$$

where τ is called the population's *mean life* (closely related to the *half-life*, $t_{1/2}$; in fact, $t_{1/2} = 0.693\tau$). A biologist suspects that a population of bacteria is decreasing exponentially as in (8.28), and measures the population on three successive days, with the results shown in the first two columns of Table 8.2. Given these data, what is his best estimate for the mean life τ?

Table 8.2. Population of bacteria.

Time t_i (days)	Population N_i	$z_i = \ln N_i$
0	153,000	11.94
1	137,000	11.83
2	128,000	11.76

If N varies as in (8.28), then the variable $z = \ln N$ should be linear in t:

$$z = \ln N = \ln N_0 - \frac{t}{\tau}. \tag{8.29}$$

Our biologist therefore calculates the three numbers $z_i = \ln N_i$ $(i = 0, 1, 2)$ shown in the third column of Table 8.2. Using these numbers, he makes a least-squares fit to the straight line (8.29) and finds as best estimates for the coefficients $\ln N_0$ and $(-1/\tau)$,

$$\ln N_0 = 11.93 \quad \text{and} \quad (-1/\tau) = -0.089 \text{ day}^{-1}.$$

The second of these implies that his best estimate for the mean life is

$$\tau = 11.2 \text{ days}.$$

The method just described is attractively simple (especially with a calculator that performs linear regression automatically) and is frequently used. Nevertheless, the method is not quite logically sound. Our derivation of the least-squares fit to a straight line $y = A + Bx$ was based on the assumption that the measured values $y_1, \ldots, y_N$ were all equally uncertain. Here we are performing our least-squares fit using the variable $z = \ln y$. Now, if the measured values y_i are all equally uncertain, then the values $z_i = \ln y_i$ are *not*. In fact, from simple error propagation we know that

$$\sigma_z = \left|\frac{dz}{dy}\right| \sigma_y = \frac{\sigma_y}{y}. \tag{8.30}$$

Thus if σ_y is the same for all measurements, then σ_z varies (with σ_z larger when y is smaller). Evidently, the variable $z = \ln y$ does not satisfy the requirement of equal uncertainties for all measurements, if y itself does.

The remedy for this difficulty is straightforward. One can modify the least-squares procedure to allow for different uncertainties in the measurements, provided the various uncertainties are known. (This method of *weighted least squares* is outlined in Problem 8.4.) If we know that the measurements of $y_1, \ldots, y_N$ really are equally uncertain, then Equation (8.30) tells us how the uncertainties in $z_1, \ldots, z_N$ vary, and we can therefore apply the method of weighted least squares to the equation $z = \ln A + Bx$.

In practice, one often cannot be sure that the uncertainties in $y_1, \ldots, y_N$ really are constant; so one can perhaps argue that one could just as well assume the uncertainties in $z_1, \ldots, z_N$ to be constant and use the simple, unweighted least squares. Often the variation in the uncertainties is small, and it makes little difference which method is used, as was true in the preceding example. In any event, straightforward application of the ordinary (unweighted) least-squares fit is an unambiguous and simple way to get *reasonable* (if not *best*) estimates for the constants A and B in the equation $y = Ae^{Bx}$; so it is frequently used in this way.

Multiple Regression

Finally, we have so far discussed only observations of *two* variables, x and y, and their relationship. In many real problems there are more than two variables to be considered. For example, in studying the pressure P of a gas, one finds that it depends on the volume V and temperature T, and one must analyze P as a function of V and T. The simplest example of such a problem is when one variable, z, depends linearly on two others, x and y:

$$z = A + Bx + Cy. \tag{8.31}$$

This problem can be analyzed by a very straightforward generalization of our two-variable method. If we have a series of measurements (x_i, y_i, z_i), $i = 1, \ldots, N$ (with the z_i all equally uncertain, and the x_i and y_i exact) then we can use the principle of maximum likelihood exactly as in Section

8.2 to show that the best estimates for the constants A, B, C are determined by normal equations of the form

$$\begin{aligned} AN + B\sum x_i + C\sum y_i &= \sum z_i, \\ A\sum x_i + B\sum x_i^2 + C\sum x_i y_i &= \sum x_i z_i, \\ A\sum y_i + B\sum x_i y_i + C\sum y_i^2 &= \sum y_i z_i. \end{aligned} \tag{8.32}$$

The equations can be solved for A, B, and C to give the best fit for the relation (8.31). This method is called *multiple regression* ("multiple" since there are more than two variables), but we will not discuss it further here.

Problems

Reminder: An asterisk indicates that the problem is discussed, or its answer given, in the Answers section at the back of the book.

***8.1** (Section 8.2). Use the method of least squares to find the line $y = A + Bx$ that best fits the four points (1, 12), (2, 13), (3, 18), (4, 19). Plot the points and line.

8.2 (Section 8.2). To find the spring constant k of a spring, a student loads it with various masses m and measures the corresponding lengths l. Her results are shown in Table 8.3.

Table 8.3.

load m (gm)	200	300	400	500	600	700	800	900
length l (cm)	5.1	5.5	5.9	6.8	7.4	7.5	8.6	9.4

Since the force mg is $k(l - l_0)$, where l_0 is the unstretched length of the spring, these data should fit the line $l = l_0 + (g/k)m$. Make a least-squares fit to the data, and find the best estimates for the unstretched length l_0 and the spring constant k.

***8.3** (Section 8.2). Suppose two variables x and y are known to satisfy the relation $y = Bx$; i.e., they lie on a straight line that is known to pass through the origin. Suppose further that you have N measurements (x_i, y_i), with the uncertainties in x negligible and those in y all equal. Using arguments like those in Section 8.2, show that the least-squares best estimate for B is

$$B = \sum x_i y_i / \sum x_i^2.$$

***8.4** (Section 8.2). Suppose we measure N pairs of values (x_i, y_i) of two variables x and y that are supposed to satisfy a linear relation $y = A + Bx$. Suppose the measurements of the x_i have negligible uncertainty, and those of the y_i have different uncertainties σ_i. (That is, y_1 has uncertainty σ_1, while y_2 has uncertainty σ_2, and so on.) Review the derivation of the least-squares fit in Section 8.2, and then generalize it to cover this situation where the uncertainties in the y_i are not all the same. Show that the best estimates of A and B are

$$A = [(\sum w_i x_i^2)(\sum w_i y_i) - (\sum w_i x_i)(\sum w_i x_i y_i)]/\Delta \tag{8.33}$$

and

$$B = [(\sum w_i)(\sum w_i x_i y_i) - (\sum w_i x_i)(\sum w_i y_i)]/\Delta, \tag{8.34}$$

with weights $w_i = 1/\sigma_i^2$ and

$$\Delta = (\sum w_i)(\sum w_i x_i^2) - (\sum w_i x_i)^2. \tag{8.35}$$

This method of *weighted least squares* can be applied only when the uncertainties σ_i (or at least their relative sizes) are known. Perhaps the commonest situation where this is so is a counting experiment, like the counting of radioactive decays. As discussed in Section 3.1 (and proved in Chapter 11), the uncertainty corresponding to any count ν is known to be $\sqrt{\nu}$.

8.5 (Section 8.2). Suppose y is known to be linear in x, so that $y = A + Bx$, and suppose we have three measurements of (x, y): $(1, 2 \pm .5)$; $(2, 3 \pm .5)$; $(3, 2 \pm 1.5)$, for which the uncertainties in x are negligible. Use the method of weighted least squares, Equations (8.33) to (8.35), to calculate A and B. Compare your results with what you would get if you ignored the variation in the uncertainties, i.e., used the unweighted fit of Equations (8.10) to (8.12). Plot the data and both lines, and try to understand the differences.

***8.6** (Section 8.4). A train, presumed to be traveling at constant speed, is timed as it goes past four different positions, with the results shown in Table 8.4. By making a least-squares fit to the line $d = d_0 + vt$, find the best estimate for the train's speed, v. What is the uncertainty in v?

Table 8.4.

distance (feet)	0	3000	6000	9000
time (seconds)	17.6	40.4	67.7	90.1

8.7 (Section 8.4). A student measures the pressure P of a gas at five different temperatures T, keeping the volume V fixed. His results are shown in Table 8.5.

Table 8.5.

pressure P_i (mm of mercury)	79	82	85	88	90
temperature T_i (°Celsius)	8	17	30	37	52

His data should fit a linear equation of the form $T = A + BP$, where A is the absolute zero of temperature (whose accepted value is $-273°$ Celsius, as discussed in Section 8.5). Find the best fit to the student's data, and hence his best estimate for absolute zero and its uncertainty.

***8.8** (Section 8.4).
(a) Use the principle of maximum likelihood, as outlined in the discussion of Equation (8.13), to show that (8.13) gives the uncertainty σ_y of y in a series of measurements $(x_1, y_1), \ldots, (x_N, y_N)$ that are supposed to fit a straight line.
(b) Use error propagation to show that the uncertainties σ_A and σ_B in the parameters of a straight line $y = A + Bx$ are given by (8.15) and (8.16).

***8.9** (Section 8.4). The least-squares fit to a set of points $(x_1, y_1), \ldots, (x_N, y_N)$ treats the variables x and y unsymmetrically. Specifically, one finds a best fit for the line $y = A + Bx$ by assuming that the numbers $y_1, \ldots, y_N$ are all equally uncertain but that $x_1, \ldots, x_N$ have negligible uncertainty. If the situation were reversed, then one would have to interchange the roles of x and y and fit to a line $x = A' + B'y$. The two lines $y = A + Bx$ and $x = A' + B'y$ would be the same if the N points lie *exactly* on a line, but in general the two lines will be slightly different. Fit the data of Problem 8.1 to a line $x = A' + B'y$ (treating the x_i as equally uncertain and the y_i as certain). Find A' and B' and their uncertainties $\sigma_{A'}$ and $\sigma_{B'}$. What would be the values of A' and B' based on the answers to Problem 8.1? Compare the lines found by the two methods. Is the difference significant?

8.10 (Section 8.6). Consider the problem of fitting a set of measurements (x_i, y_i), $i = 1, \ldots, N$, to the polynomial $y = A + Bx + Cx^2$. Use the principle of maximum likelihood to show that the best estimates for A, B, C based on the data are given by Equations (8.23). Follow the arguments outlined between Equations (8.20) and (8.23).

***8.11** (Section 8.6). One way to measure the acceleration of a freely falling body is to measure its height y_i at a succession of equally spaced times t_i (with a multiflash photograph, for example) and to find the best fit to the expected polynomial

$$y = y_0 + v_0 t - \tfrac{1}{2} g t^2. \tag{8.36}$$

Use the equations (8.23) to find the best estimates for the three coefficients in (8.36), and hence the best estimate for g, based on the five measurements in Table 8.6.

Table 8.6.

time t (tenths of sec)	-2	-1	0	1	2
height y (cm)	131	113	89	51	7

Note that we can name the times however we like. A more natural choice might seem to be $t = 0, 1, \ldots, 4$. However, when you solve the problem you will see that defining the times to be symmetrically spaced about zero causes about half the sums involved to be zero and greatly simplifies the calculations. This trick can be used whenever the values of the independent variable are equally spaced.

8.12 (Section 8.6). Suppose that y is expected to have the form $y = Af(x) + Bg(x)$, where A and B are unknown parameters, and f and g are fixed, known functions (such as $f = x$ and $g = x^2$, or $f = \cos x$ and $g = \sin x$). Use the principle of maximum likelihood to show that the best estimates for A and B, based on data (x_i, y_i), $i = 1, \ldots, N$, must satisfy

$$\begin{aligned} A\sum[f(x_i)]^2 + B\sum f(x_i)g(x_i) &= \sum y_i f(x_i), \\ A\sum f(x_i)g(x_i) + B\sum[g(x_i)]^2 &= \sum y_i g(x_i). \end{aligned} \tag{8.37}$$

***8.13** (Section 8.6). A weight oscillating on a vertical spring should have height y given by

$$y = A \cos \omega t + B \sin \omega t.$$

A student measures ω to be 10 rad/sec with negligible uncertainty. Using a multiflash photograph, she then finds y for five equally spaced times, as shown in Table 8.7.

Table 8.7.

t (tenths of sec)	-4	-2	0	2	4
y (cm)	3	-16	6	9	-8

Use Equations (8.37) to find best estimates for A and B. Plot the data and your best fit. (If you plot the data first, you will have the opportunity to consider how hard it would be to choose a best fit without the least-squares method.) If the student judges that her measured values of y were uncertain by "a couple of centimeters," would you say that the data are an acceptable fit to the expected curve?

***8.14** (Section 8.6). The rate R at which a sample of radioactive material emits radiation decreases exponentially as the material is depleted:

$$R = R_0 e^{-t/\tau},$$

where τ is the mean life of the sample. A student observed a certain radioactive material for three hours with the results shown in Table 8.8. By making a least-squares fit to the line $\ln R = \ln R_0 - t/\tau$, find the best estimate for the mean life τ.

Table 8.8.

time t (hours)	0	1	2	3
rate R (arbitrary units)	13.8	7.9	6.1	2.9

CHAPTER 9

Covariance and Correlation

In this chapter we introduce the important concept of covariance. The notion of covariance arises naturally in the discussion of error propagation; so we will introduce it in Section 9.2 after briefly reviewing error propagation in Section 9.1. In Section 9.3 we will then use the covariance to define the coefficient of linear correlation for N measured points (x_1, y_1), $\ldots$, (x_N, y_N). This coefficient, denoted r, provides a measure of how well the observed points (x_i, y_i) fi a straight line of the form $y = A + Bx$. Its use is discussed in Sections 9.4 and 9.5.

9.1. Review of Error Propagation

In this and the next section we return for a final look at the important question of error propagation. We first discussed error propagation in Chapter 3, where we reached several conclusions. We imagined measuring two quantities x and y in order to calculate some function $q(x, y)$, such as $q = x + y$ or $q = x^2 \sin y$. (In fact, we discussed there a function $q(x, \ldots, z)$ of an arbitrary number of variables $x, \ldots, z$; for simplicity we will now consider just two variables.) A simple argument suggested that the uncertainty in our answer for q is just

$$\delta q \approx \left|\frac{\partial q}{\partial x}\right| \delta x + \left|\frac{\partial q}{\partial y}\right| \delta y. \tag{9.1}$$

We first derived this for the simple special cases of sums, differences, products, and quotients. For instance, if q is the sum $q = x + y$, then (9.1) reduces to the familiar $\delta q \approx \delta x + \delta y$. The general result (9.1) was derived in Equation (3.43).

We next recognized that (9.1) is often probably an overstatement of δq, since there may be partial cancellation of the errors in x and y. We stated, without proof, that when the errors in x and y are independent and random, a better value for the uncertainty in the calculated value of $q(x, y)$ is

$$\delta q = \sqrt{\left(\frac{\partial q}{\partial x}\,\delta x\right)^2 + \left(\frac{\partial q}{\partial y}\,\delta y\right)^2}. \tag{9.2}$$

We also stated, without proof, that whether or not the errors are independent and random, the simpler formula (9.1) always gives an upper bound on δq; that is, the uncertainty δq is never any worse than is given by (9.1).

In Chapter 5 we gave a proper definition and proof of (9.2). First, we saw that a good measure of the uncertainty δx in a measurement is given by the standard deviation σ_x; in particular, we saw that if the measurements of x are normally distributed, then we can be 68 percent confident that the measured value lies within σ_x of the true value. Second, we saw that if the measurements of x and y are governed by independent normal distributions, with standard deviations σ_x and σ_y, then the values of $q(x, y)$ are also normally distributed, with standard deviation

$$\sigma_q = \sqrt{\left(\frac{\partial q}{\partial x}\,\sigma_x\right)^2 + \left(\frac{\partial q}{\partial y}\,\sigma_y\right)^2}. \tag{9.3}$$

This result provides the justification for our claim in (9.2).

In Section 9.2 we will derive a precise formula for the uncertainty in q that applies whether or not the errors in x and y are independent and normally distributed. In particular, we will prove that (9.1) always provides an upper bound on the uncertainty in q.

Before we derive these results, let us first review the definition of the standard deviation. The standard deviation σ_x of N measurements $x_1, \ldots, x_N$ was originally defined by the equation

$$\sigma_x{}^2 = \frac{1}{N}\sum_{i=1}^{N}(x_i - \bar{x})^2. \tag{9.4}$$

If the measurements of x are normally distributed, then in the limit that N is large, the definition (9.4) is equivalent to defining σ_x as the width parameter that appears in the Gauss function

$$\frac{1}{\sigma_x\sqrt{2\pi}}\,e^{-(x-X)^2/2\sigma_x{}^2}$$

that governs the measurements of x. Since we will now be considering the possibility that the errors in x may not be normally distributed, this second definition is no longer available to us. However, we can, and will, still define σ_x by (9.4). Whether or not the distribution of errors is normal, this definition of σ_x gives a reasonable measure of the random uncertainties in our measurement of x. (As in Chapter 5, we will suppose that all systematic errors have been identified and reduced to a negligible level, so that all remaining errors are random.)

There remains the usual ambiguity as to whether we use the definition (9.4) of σ_x, or the "improved" definition with the factor N in the denominator replaced by $(N - 1)$. Fortunately, the discussion that follows applies to either definition, as long as we are consistent in our use of one or the other. For convenience, we will use the definition (9.4), with N in the denominator.

9.2. Covariance in Error Propagation

Suppose that, in order to find a value for the function $q(x, y)$, we measure the two quantities x and y several times, obtaining N pairs of data, $(x_1, y_1), \ldots, (x_N, y_N)$. From the N measurements $x_1, \ldots, x_N$, we can compute the mean $\bar{x}$ and standard deviation σ_x in the usual way; similarly, from $y_1, \ldots, y_N$, we can compute $\bar{y}$ and σ_y. Next, using the N pairs of measurements we can compute N values of the quantity of interest

$$q_i = q(x_i, y_i), \qquad (i = 1, \ldots, N).$$

Given $q_1, \ldots, q_N$, we can now calculate their mean $\bar{q}$, which we assume gives our best estimate for q, and their standard deviation σ_q, which is our measure of the random uncertainty in the values q_i.

We will assume, as usual, that all our uncertainties are small, and hence that all the numbers $x_1, \ldots, x_N$ are close to $\bar{x}$ and that all the $y_1, \ldots, y_N$ are close to $\bar{y}$. We can then make the approximation

$$\begin{aligned} q_i &= q(x_i, y_i) \\ &\approx q(\bar{x}, \bar{y}) + \frac{\partial q}{\partial x}(x_i - \bar{x}) + \frac{\partial q}{\partial y}(y_i - \bar{y}). \end{aligned} \tag{9.5}$$

In this expression the partial derivatives $\partial q/\partial x$ and $\partial q/\partial y$ are taken at the point $x = \bar{x}$, $y = \bar{y}$, and are therefore the same for all $i = 1, \ldots, N$. With this approximation, the mean becomes

$$\begin{aligned}\bar{q} &= \frac{1}{N}\sum_{i=1}^{N} q_i \\ &= \frac{1}{N}\sum_{i=1}^{N}\left[q(\bar{x},\bar{y}) + \frac{\partial q}{\partial x}(x_i - \bar{x}) + \frac{\partial q}{\partial y}(y_i - \bar{y})\right].\end{aligned}$$

This gives $\bar{q}$ as the sum of three terms. The first term is just $q(\bar{x}, \bar{y})$, and the other two are exactly zero. (For example, it follows from the definition of $\bar{x}$ that $\sum(x_i - \bar{x}) = 0$.) Thus we have the remarkably simple result

$$\bar{q} = q(\bar{x}, \bar{y}); \tag{9.6}$$

that is, to find the mean $\bar{q}$ we have only to calculate the function $q(x, y)$ at the point $x = \bar{x}$ and $y = \bar{y}$.

The standard deviation in the N values $q_1, \ldots, q_N$ is given by

$$\sigma_q{}^2 = \frac{1}{N}\sum(q_i - \bar{q})^2.$$

Substituting (9.5) and (9.6), we find that

$$\begin{aligned}\sigma_q{}^2 &= \frac{1}{N}\sum\left[\frac{\partial q}{\partial x}(x_i - \bar{x}) + \frac{\partial q}{\partial y}(y_i - \bar{y})\right]^2 \\ &= \left(\frac{\partial q}{\partial x}\right)^2 \frac{1}{N}\sum(x_i - \bar{x})^2 + \left(\frac{\partial q}{\partial y}\right)^2 \frac{1}{N}\sum(y_i - \bar{y})^2 \\ &\quad + 2\frac{\partial q}{\partial x}\frac{\partial q}{\partial y}\frac{1}{N}\sum(x_i - \bar{x})(y_i - \bar{y}).\end{aligned} \tag{9.7}$$

The sums in the first two terms are those that appear in the definition of the standard deviations σ_x and σ_y. The final sum is one that we have not encountered before. It is called the *covariance*[1] of x and y, and is denoted

$$\sigma_{xy} = \frac{1}{N}\sum_{i=1}^{N}(x_i - \bar{x})(y_i - \bar{y}). \tag{9.8}$$

[1] The name *covariance* for σ_{xy} (for two variables x, y) parallels the name *variance* for $\sigma_x{}^2$ (for one variable x). To emphasize this parallel, the covariance (9.8) is sometimes denoted $\sigma_{xy}{}^2$, not an especially happy notation, since the covariance can be negative. A convenient feature of our definition (9.8) is that σ_{xy} has the dimensions of xy, just as σ_x has the dimensions of x.

With this definition, the equation (9.7) for the standard deviation σ_q becomes

$$\sigma_q{}^2 = \left(\frac{\partial q}{\partial x}\right)^2 \sigma_x{}^2 + \left(\frac{\partial q}{\partial y}\right)^2 \sigma_y{}^2 + 2\frac{\partial q}{\partial x}\frac{\partial q}{\partial y}\sigma_{xy}. \tag{9.9}$$

This gives the standard deviation σ_q, whether or not the measurements of x and y are independent, and whether or not they are normally distributed.

If the measurements of x and y *are* independent, it is easily seen that, after many measurements, the covariance σ_{xy} should approach zero. Whatever the value of y_i, the quantity $x_i - \bar{x}$ is just as likely to be negative as it is to be positive. Thus after many measurements the positive and negative terms in (9.8) should nearly balance; and in the limit of infinitely many measurements, the factor $1/N$ in (9.8) guarantees that σ_{xy} is zero. (After a finite number of measurements, σ_{xy} will not be exactly zero, but it should be *small*, if the errors in x and y really are independent and random.) With σ_{xy} zero, Equation (9.9) for σ_q reduces to

$$\sigma_q{}^2 = \left(\frac{\partial q}{\partial x}\right)^2 \sigma_x{}^2 + \left(\frac{\partial q}{\partial y}\right)^2 \sigma_y{}^2, \tag{9.10}$$

the familiar result for independent and random uncertainties.

If the measurements of x and y are *not* independent, then the covariance σ_{xy} need not be zero. For instance, it is easy to imagine a situation where an overestimate of x will always be accompanied by an overestimate of y, and vice versa. The numbers $(x_i - \bar{x})$ and $(y_i - \bar{y})$ will then always have the same sign (both positive or both negative), and their product will always be positive. Since all terms in the sum (9.8) are positive, σ_{xy} need not vanish, even in the limit that we make infinitely many measurements.

When the covariance σ_{xy} is not zero (even in the limit of infinitely many measurements), we say that the errors in x and y are *correlated.* In this situation the uncertainty σ_q in $q(x, y)$ as given by (9.9) is *not* the same as we would get from the formula (9.10) for independent and random errors.

Using the formula (9.9), we can derive an upper limit on σ_q that is always valid. It is a simple algebraic exercise (Problem 9.1) to prove that the covariance σ_{xy} satisfies the so-called *Schwarz inequality*

$$|\sigma_{xy}| \leqslant \sigma_x \sigma_y. \tag{9.11}$$

If we substitute (9.11) into the expression (9.9) for the uncertainty σ_q, we find that

$$\sigma_q{}^2 \leqslant \left(\frac{\partial q}{\partial x}\right)^2 \sigma_x{}^2 + \left(\frac{\partial q}{\partial y}\right)^2 \sigma_y{}^2 + 2\left|\frac{\partial q}{\partial x}\frac{\partial q}{\partial y}\right| \sigma_x \sigma_y$$
$$= \left[\left|\frac{\partial q}{\partial x}\right| \sigma_x + \left|\frac{\partial q}{\partial y}\right| \sigma_y\right]^2;$$

that is,

$$\sigma_q \leqslant \left|\frac{\partial q}{\partial x}\right| \sigma_x + \left|\frac{\partial q}{\partial y}\right| \sigma_y. \tag{9.12}$$

With this result we have finally established the precise significance of our original, simple expression

$$\delta q \approx \left|\frac{\partial q}{\partial x}\right| \delta x + \left|\frac{\partial q}{\partial y}\right| \delta y \tag{9.13}$$

for the uncertainty δq. If we adopt the standard deviation σ_q as our measure of the uncertainty in q, then (9.12) shows that the old expression (9.13) is really the *upper limit* on the uncertainty. Whether or not the errors in x and y are independent, and whether or not they are normally distributed, the uncertainty in q will never exceed the right side of (9.13). If the measurements of x and y are correlated in just such a way that $|\sigma_{xy}| = \sigma_x \sigma_y$, its largest possible value according to (9.11), then the uncertainty in q can actually be as large as given by (9.13), but it can never be any larger.

The role of the covariance σ_{xy} in this discussion of error propagation is purely theoretical; and, in fact, the concept of the covariance does not often play a practical role in error propagation (at least in the elementary physics laboratory). We next discuss a problem in which it does play a central and practical role.

9.3. *Coefficient of Linear Correlation*

The notion of covariance σ_{xy} introduced in Section 9.2 enables us to answer a question raised in Chapter 8, how well a set of measurements $(x_1, y_1), \ldots, (x_N, y_N)$ of two variables supports the hypothesis that x and y are linearly related.

Let us suppose we have measured N pairs of values $(x_1, y_1), \ldots, (x_N, y_N)$ of two variables that we suspect should satisfy a linear relation of the form

$$y = A + Bx.$$

It is important to note that $x_1, \ldots, x_N$ are no longer measurements of one single number, as they were in the last two sections; rather, they are measurements of N different values of some variable (for example, N different heights from which we have dropped a stone). The same applies to $y_1, \ldots, y_N$.

Using the method of least squares, we can find the values of A and B for the line that best fits the points $(x_1, y_1), \ldots, (x_N, y_N)$. If we already have a reliable estimate of the uncertainties in the measurements, then we can see whether the measured points do lie reasonably close to the line (compared to the known uncertainties). If they do, then the measurements support our suspicion that x and y are linearly related.

Unfortunately, in many experiments it is hard to get a reliable estimate of the uncertainties in advance, and we must use the data themselves to decide whether the two variables appear to be linearly related. In particular, there is a type of experiment where it is *impossible* to know the size of uncertainties in advance. This type of experiment, which is more common in the social than the physical sciences, is best explained by an example.

Suppose a professor, anxious to convince his students that doing homework will help them to do well in exams, keeps records of their scores on homeworks and exams, and plots them on a "scatter plot" as in Figure 9.1. In this figure, homework scores are plotted horizontally and exam scores vertically. Each point (x_i, y_i) shows one student's homework score, x_i, and exam score, y_i. What the professor hopes to show is that high exam scores tend to be *correlated* with high homework scores, and vice versa (and his scatter plot certainly suggests that this is approximately so). In this kind of experiment, there are no uncertainties in the points; each student's two scores are known exactly. The uncertainty lies rather in the extent to which the scores *are correlated*; and this has to be decided from the data.

The two variables x and y (in either a typical physics experiment or one like that just described) may, of course, be related by a more complicated relation than the simple linear one, $y = A + Bx$. For example, plenty of physical laws lead to quadratic relations of the form $y = A + Bx + Cx^2$. Nevertheless, we will restrict our discussion here to the simpler problem of deciding whether a given set of points supports the hypothesis of a *linear* relation $y = A + Bx$.

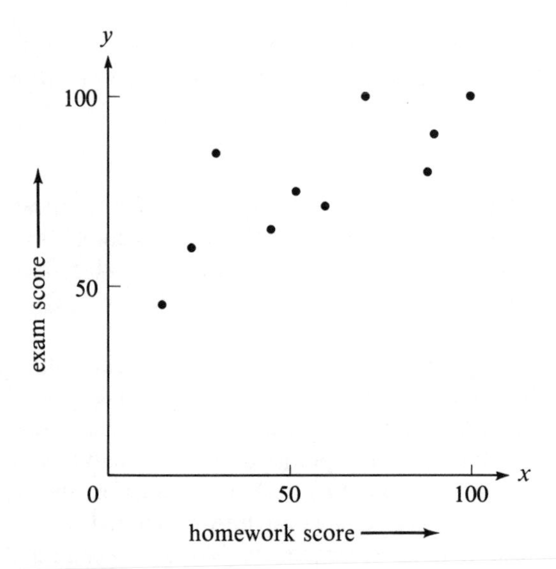

Figure 9.1. A "scatter plot" showing students' scores on exams and homeworks. Each of the ten points (x_i, y_i) shows a student's homework score, x_i, and exam score, y_i.

The extent to which a set of points $(x_1, y_1), \ldots, (x_N, y_N)$ support a linear relation between x and y is measured by the *linear correlation coefficient*, or just *correlation coefficient*,

$$r = \frac{\sigma_{xy}}{\sigma_x \sigma_y}, \tag{9.14}$$

where the covariance σ_{xy} and standard deviations σ_x and σ_y are defined exactly as before, in Equations (9.8) and (9.4).[2] Substituting these definitions into (9.14), we can rewrite the correlation coefficient as

$$r = \frac{\sum(x_i - \bar{x})(y_i - \bar{y})}{[\sum(x_i - \bar{x})^2 \sum(y_i - \bar{y})^2]^{1/2}}. \tag{9.15}$$

[2] Notice, however, that their significance is slightly different. For example, in Section 9.2 $x_1, \ldots, x_N$ were measurements of *one number*, and if these measurements were precise, σ_x should be small. In the present case $x_1, \ldots, x_N$ are measurements of *different* values of a variable, and even if the measurements are precise, there is no reason to think σ_x will be small. Note also that some authors use the number r^2, called the *coefficient of determination.*

As we will soon show, the number r is an indicator of how well the points (x_i, y_i) fit a straight line. It is a number somewhere between -1 and 1. If r is close to ± 1, then the points lie close to some straight line; if r is close to 0, then the points are uncorrelated, with little or no tendency to lie on a straight line.

To prove these assertions, we first observe that the Schwarz inequality (9.11), $|\sigma_{xy}| \leqslant \sigma_x \sigma_y$, implies immediately that $|r| \leqslant 1$ or

$$-1 \leqslant r \leqslant 1$$

as claimed. Next, let us suppose that the points (x_i, y_i) all lie *exactly* on the line $y = A + Bx$. In this case $y_i = A + Bx_i$ for all i, and hence $\bar{y} = A + B\bar{x}$. Subtracting these two equations, we see that

$$y_i - \bar{y} = B(x_i - \bar{x})$$

for each i. Inserting these into (9.15), we find that

$$r = \frac{B \sum (x_i - \bar{x})^2}{[\sum (x_i - \bar{x})^2 B^2 \sum (x_i - \bar{x})^2]^{1/2}} = \frac{B}{|B|} = \pm 1. \tag{9.16}$$

That is, if the points $(x_1, y_1), \ldots, (x_N, y_N)$ lie perfectly on a line, then $r = \pm 1$, the sign of r being determined by the slope of the line ($r = 1$ for B positive, and $r = -1$ for B negative).[3] Even when the variables x and y really are linearly related, we do not expect our experimental points to lie *exactly* on a line. Thus we do not expect r to be exactly ± 1. On the other hand, we do expect a value of r that is *close to* ± 1, if we believe that x and y are linearly related.

Suppose, on the other hand, that there is no relationship between the variables x and y. Whatever the value of y_i, each x_i would then be just as likely to be above $\bar{x}$ as below $\bar{x}$. Thus the terms in the sum

$$\sum (x_i - \bar{x})(y_i - \bar{y})$$

in the numerator of r in (9.15) are just as likely to be positive as negative. Meanwhile, the terms in the denominator of r are all positive. Thus, in the limit that N, the number of measurements, approaches infinity, the correlation coefficient r will be zero. With a finite number of data points,

[3] If the line is exactly horizontal, then $B = 0$, and (9.16) gives $r = 0/0$; i.e., r is undefined. Fortunately, this special case is not important in practice, since it corresponds to y being a constant, independent of x.

we do not expect r to be exactly zero, but we do expect it to be *small* (if the two variables really are unrelated).

If two variables x and y are such that, in the limit of infinitely many measurements, their covariance σ_{xy} is zero (and hence $r = 0$), we say that the variables are *uncorrelated.* If, after a finite number of measurements, the correlation coefficient $r = \sigma_{xy}/\sigma_x\sigma_y$ is small, then this supports the hypothesis that x and y are uncorrelated.

As an example, we can consider the exam and homework scores shown in Figure 9.1. These scores are given in Table 9.1. A simple calculation (Problem 9.4) shows that the correlation coefficient for these ten pairs of scores is $r = 0.8$. The professor concludes that this is "reasonably close" to 1, and so can announce to next year's class that, since there is a good correlation between homework and exam scores, it is important to do the homework.

Table 9.1. Students' scores.

Student, i	1	2	3	4	5	6	7	8	9	10
Homework, x_i	90	60	45	100	15	23	52	30	71	88
Exam, y_i	90	71	65	100	45	60	75	85	100	80

If our professor had found a correlation coefficient r close to zero, then he would have been in the embarrassing position of having shown that homework scores have no bearing on exam scores. If r had turned out to be close to -1, then he would have made the even more embarrassing discovery that homework and exam scores show a *negative* correlation; that is, that students who do a good job on homework tend to do poorly on the exam.

9.4. *Quantitative Significance of r*

It should be clear from the example just discussed that we do not yet have a complete answer to our original question about how well data points support a linear relation between x and y. Our professor found a correlation coefficient $r = .8$, and judged this "reasonably close" to 1. But how can we decide objectively what is "reasonably close to 1"? Would $r = .6$ have been reasonably close? Or $r = .4$? We can answer these questions by the following argument.

Suppose the two variables x and y are in reality *uncorrelated*; that is, in the limit of infinitely many measurements, the correlation coefficient r would be zero. After a finite number of measurements, r is very unlikely to be exactly zero. One can, in fact, calculate the probability that r will exceed any specific value. We shall denote by

$$P_N(|r| \geqslant r_o)$$

the probability that N measurements of two uncorrelated variables x and y will give a coefficient r larger[4] than any particular r_o. For instance, we could calculate the probability

$$P_N(|r| \geqslant 0.8)$$

that, after N measurements of the uncorrelated variables x and y, the correlation coefficient would be at least as large as our professor's 0.8. The calculation of these probabilities is quite complicated, and will not be given here. However, the results for a few representative values of the parameters are shown in Table 9.2, and a more complete tabulation is given in Appendix C.

Table 9.2. The probability $P_N(|r| \geqslant r_o)$ that N measurements of two uncorrelated variables x and y would produce a correlation coefficient with $|r| \geqslant r_o$. Values given are percentage probabilities, and blanks indicate values less than .05 percent.

	r_o										
N	0	.1	.2	.3	.4	.5	.6	.7	.8	.9	1
3	100	94	87	81	74	67	59	51	41	29	0
6	100	85	70	56	43	31	21	12	6	1	0
10	100	78	58	40	25	14	7	2	.5	—	0
20	100	67	40	20	8	2	.5	.1	—	—	0
50	100	49	16	3	.4	—	—	—	—	—	0

Although we have not shown how the probabilities in Table 9.2 are calculated, we can understand their general behavior and put them to use. The left-hand column shows the number of data points N. (In our example, the professor recorded ten students' scores; so $N = 10$.) The numbers

[4] Since a correlation is indicated if r is close to $+1$ *or* to -1, we consider the probability of getting the *absolute value* $|r| \geqslant r_o$.

in each succeeding column show the probability that N measurements of two *uncorrelated* variables would yield a coefficient r at least as big as the number at the top of the column. For example, we see that the probability that ten uncorrelated data points would give $|r| \geqslant 0.8$ is only .5 percent, not a large probability. Our professor can therefore say it is *very unlikely* that uncorrelated scores would have produced a coefficient with $|r|$ greater than or equal to the 0.8 that he obtained. In other words, it is very *likely* that the scores on homeworks and exams really are correlated.

Several features of Table 9.2 deserve comment. All entries in the first column are 100 percent, because $|r|$ is always greater than or equal to zero; so the probability of finding $|r| \geqslant 0$ is always 100 percent. Similarly the entries in the last column are all zero, since the probability of finding $|r| \geqslant 1$ is zero.[5] The numbers in the intermediate columns vary with the number of data points N. This also is easily understood. If we make just three measurements, the chance of their having a correlation coefficient with $|r| \geqslant 0.5$, say, is obviously quite good (67 percent, in fact); but if we make 20 measurements, and the two variables really are uncorrelated, the chance of finding $|r| \geqslant 0.5$ is obviously very small (2 percent, in fact).

Armed with the probabilities in Table 9.2 (or in the more complete table in Appendix C), we now have the most complete possible answer to the question of how well N pairs of values (x_i, y_i) support a linear relation between x and y. From the measured points, we can first calculate the observed correlation coefficient r_o (the subscript o stands for "observed"). Next, using one of these tables, we can find the probability $P_N(|r| \geqslant |r_o|)$ that N uncorrelated points would have given a coefficient at least as large as the observed coefficient r_o. If this probability is "sufficiently small," then we conclude that it is very *improbable* that x and y are uncorrelated, and hence very *probable* that they really are correlated.

We still have to choose the value of the probability that we will regard as "sufficiently small." One fairly common choice is to regard an observed correlation r_o as "significant" if the probability of obtaining a coefficient r with $|r| \geqslant |r_o|$ from uncorrelated variables is less than 5 percent. A correlation is sometimes called "highly significant" if the corresponding probability is less than 1 percent. Whatever choice we make, we do *not* get a definite answer that the data are, or are not, correlated; instead, we have a quantitative measure of how improbable it is that they are uncorrelated.

[5] Although it is *impossible* that $|r| > 1$, it is, in principle, possible that $|r| = 1$. However, r is a continuous variable, and the probability of getting $|r|$ exactly equal to one is zero. Thus $P_N(|r| \geqslant 1) = 0$.

9.5. Examples

Suppose we measure three pairs of values (x_i, y_i) and find that they have a correlation coefficient of 0.7 (or -0.7). Does this support the hypothesis that x and y are linearly related?

Referring to Table 9.2, we see that even if the variables x and y were completely uncorrelated, the probability is 51 percent for getting $|r| \geqslant 0.7$ when $N = 3$. In other words, it is entirely possible that x and y are uncorrelated; so we have no worthwhile evidence of correlation. In fact, with only three measurements it would be very difficult to get convincing evidence of a correlation. Even an observed coefficient as large as 0.9 is quite insufficient, since the probability is 29 percent for getting $|r| \geqslant 0.9$ from three measurements of uncorrelated variables.

If we found a correlation of 0.7 from six measurements, the situation would be a little better, but still not good enough. With $N = 6$, the probability of getting $|r| \geqslant 0.7$ from uncorrelated variables is 12 percent. This is not small enough to rule out the possibility that x and y are uncorrelated.

On the other hand, if we found $r = 0.7$ after 20 measurements, then we would have strong evidence for a correlation, since when $N = 20$ the probability of getting $|r| \geqslant 0.7$ from two uncorrelated variables is only 0.1 percent. By any standards this is very improbable, and we could confidently argue that a correlation is indicated. In particular, the correlation could be called "highly significant," since the probability concerned is less than 1 percent.

Problems

Reminder: An asterisk indicates that the problem is discussed, or its answer given, in the Answers section at the back of the book.

***9.1** (Section 9.2). Prove that the covariance σ_{xy}, defined in (9.8), satisfies the Schwarz inequality (9.11),

$$|\sigma_{xy}| \leqslant \sigma_x \sigma_y. \tag{9.17}$$

Hint: let t be an arbitrary number and consider the function

$$A(t) = \frac{1}{N} \sum [(x_i - \bar{x}) + t(y_i - \bar{y})]^2 \geqslant 0. \tag{9.18}$$

Since $A(t)$ is positive whatever the value of t, you can find its minimum value A_{min} by setting its derivative dA/dt equal to zero, and this A_{min} is still greater than or equal to zero. Show that $A_{\text{min}} = \sigma_x^2 - (\sigma_{xy}^2/\sigma_y^2)$, and deduce (9.17).

9.2 (Section 9.2).
(a) Imagine a series of N measurements of two fixed lengths x and y, made to find a value for some function $q(x, y)$. Suppose that several different tape measures are used, but that each pair (x_i, y_i) is measured with the same tape; i.e., the pair (x_1, y_1) is measured with one tape, (x_2, y_2) is measured with one tape, and so on. Assuming that the main source of errors is that some of the tapes have shrunk and some stretched, show clearly that the covariance σ_{xy} is bound to be positive.
(b) Show further, under the same conditions, that $\sigma_{xy} = \sigma_x \sigma_y$, that is, that σ_{xy} is as large as permitted by the Schwarz inequality (9.17). *Hint*: assume the ith tape has shrunk by a factor α_i (with α_i close to 1); then a length that is really X will be measured as $x_i = \alpha_i X$. The moral of this problem is that there are circumstances where the covariance is certainly not negligible.

***9.3** (Section 9.3).
(a) Prove the identity

$$\sum(x_i - \bar{x})(y_i - \bar{y}) = \sum x_i y_i - N\bar{x}\bar{y}.$$

(b) Hence show that the correlation coefficient r defined in (9.15) can be written as

$$r = \frac{\sum x_i y_i - N\bar{x}\bar{y}}{[(\sum x_i^2 - N\bar{x}^2)(\sum y_i^2 - N\bar{y}^2)]^{1/2}}. \tag{9.19}$$

This is often a more convenient way to calculate r, since it avoids the need to compute the individual deviations $x_i - \bar{x}$ and $y_i - \bar{y}$.

9.4 (Section 9.4).
(a) Check that the correlation coefficient r for the ten pairs of scores in Table 9.1 is $r \approx 0.8$.
(b) Using the table of probabilities in Appendix C, find the probability that one would obtain a correlation r with $|r| \geqslant 0.8$ if the two scores were really uncorrelated.

***9.5** (Section 9.4). In the photoelectric effect, the kinetic energy K of ejected electrons is supposed to be a linear function of the frequency f of the light used,

$$K = hf - \phi, \tag{9.20}$$

where h and ϕ are constants. To check this, a student measures K for N different values of f and calculates the correlation coefficient r for her results.

(a) If she makes five measurements ($N = 5$) and finds $r = .7$, does she have significant support for the linear relation (9.20)?

(b) What if $N = 20$ and $r = .5$?

***9.6** (Section 9.4).

(a) Draw a scatter plot for these five pairs of measurements:

$$x = 1 \quad 2 \quad 3 \quad 4 \quad 5$$
$$y = 4 \quad 4 \quad 3 \quad 2 \quad 1$$

Compute their correlation coefficient r. It is probably simpler to use the form (9.19). Do the data show a significant correlation? The necessary probabilities can be found in Appendix C.

(b) Repeat part (a) for the following data:

$$x = 1 \quad 2 \quad 3 \quad 4 \quad 5$$
$$y = 3 \quad 1 \quad 2 \quad 2 \quad 1$$

9.7 (Section 9.4). A psychologist, investigating the relation between the intelligence of fathers and sons, measures I.Q.s for ten fathers and their sons, with the results shown in Table 9.3, where x_i = father's I.Q., y_i = corresponding son's I.Q.

Table 9.3.

x_i	74	83	85	96	98	100	106	107	120	124
y_i	76	103	99	109	111	107	91	101	120	119

Do these data support a correlation between the intelligence of fathers and sons?

CHAPTER 10

The Binomial Distribution

The Gauss, or normal, distribution is the only example of a distribution that we have studied so far. We will now discuss two other important examples, the binomial distribution (in this chapter) and the Poisson distribution (in Chapter 11).

10.1. Distributions

In Chapter 5 we introduced the idea of a *distribution*, the function that describes the proportion of times that a repeated measurement yields each of its various possible answers. For example, we could make N measurements of the period T of a pendulum and find the distribution of our various measured values of T; or we could measure the heights h of N Americans and find the distribution of the various measured heights h.

We next introduced the notion of the *limiting distribution*, the distribution that would be obtained in the limit that the number of measurements N becomes very large. The limiting distribution can be viewed as telling us the *probability* that one measurement will yield any of the possible values: the probability that one measurement of the period will yield any particular value T; the probability that one American (chosen at random) will have any particular height h. For this reason the limiting distribution is also sometimes called the *probability distribution*.

Of the many possible limiting distributions, the only one we have discussed is the Gauss, or normal, distribution, which describes the distribution of answers for any measurement subject to many sources of error that are all random and small. As such, the Gauss distribution is the most important of all limiting distributions for the physical scientist, and amply deserves the prominence given it here. Nevertheless, there are several other distributions of great theoretical or practical importance, and in this and the next chapter we give two examples.

In this chapter we shall describe the binomial distribution. This distribution is not of great practical importance to the experimental physicist. However, its simplicity makes it an excellent introduction to many properties of distributions, and it is theoretically important, since from it we can derive the all-important Gauss distribution.

10.2. Probabilities in Dice Throwing

The binomial distribution can best be described by an example. Suppose we undertake as our "experiment" to throw three dice and to measure the number of aces showing. The possible results of the experiment are the answers 0, 1, 2, or 3 aces. If we repeat the experiment an enormous number of times, then we will find the limiting distribution, which will tell us the probability that in any one throw (of all three dice) we get ν aces, where $\nu = 0, 1, 2$, or 3.

This experiment is sufficiently simple that we can easily calculate the probability of the four possible outcomes. We observe first that, assuming the dice are true, the probability of getting an ace when throwing *one* die is $\frac{1}{6}$. Let us now throw all three dice, and ask first for the probability of getting three aces ($\nu = 3$). Since each separate die has probability $\frac{1}{6}$ of showing an ace, and since the three dice roll independently, the probability for three aces is

$$P(3 \text{ aces in } 3 \text{ throws}) = (\tfrac{1}{6})^3$$
$$\approx 0.5\%.$$

To calculate the probability for two aces ($\nu = 2$) is a little harder, since we can throw two aces in several ways. The first and second dice could show aces and the third not (A, A, not A), or the first and third could show aces and the second not (A, not A, A), and so on. Here we argue in two steps. First, we consider the probability of throwing two aces in any definite order, such as (A, A, not A). The probability that the first die will show an ace is $\frac{1}{6}$, and likewise for the second. On the other hand, the probability that the last die will *not* show an ace is $\frac{5}{6}$. Thus the probability for two aces in this particular order is

$$P(A, A, \text{not } A) = (\tfrac{1}{6})^2 \times (\tfrac{5}{6}).$$

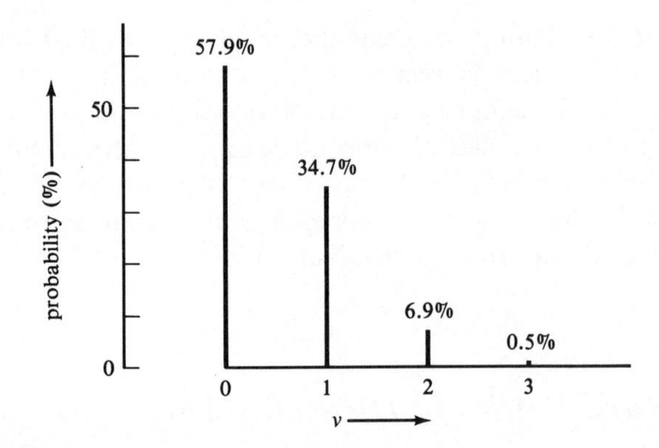

Figure 10.1. Probability of getting ν aces when throwing three dice. This function is the binomial distribution $b_{n,p}(\nu)$, with $n = 3$ and $p = \frac{1}{6}$.

The probability for two aces in any other definite order is the same. Finally, there are three different orders in which we could get our two aces: (A, A, not A), or (A, not A, A), or (not A, A, A). Thus the total probability for getting two aces (in any order) is

$$\begin{aligned} P(2 \text{ aces in 3 throws}) &= 3 \times (\tfrac{1}{6})^2 \times (\tfrac{5}{6}) \\ &\approx 6.9\%. \end{aligned} \tag{10.1}$$

Similar calculations give the probabilities for one ace in three throws (34.7 percent), and for no aces in three throws (57.9 percent). Our numerical conclusions can be summarized by drawing the probability distribution for the number of aces obtained when throwing three dice, as in Figure 10.1. This distribution is an example of the binomial distribution, the general form of which we now describe.

10.3. Definition of the Binomial Distribution

To describe the general binomial distribution, we need to introduce some terminology. First, we imagine making n independent *trials*, such as throwing n dice, tossing n coins, or testing n firecrackers. Each trial can have various outcomes; a die can show any face from 1 to 6, a coin can show heads or tails, a firecracker can explode or fizzle. We refer to the outcome in which we happen to be interested as a *success*. Thus "success" could be throwing an ace on a die, or a head on a coin, or having a fire-cracker explode. We denote by p the probability of success in any one trial, and by $q = 1 - p$ that of "failure" (i.e., of getting any outcome other than the one of interest). Thus, $p = \frac{1}{6}$ for getting an ace on a die, $p = \frac{1}{2}$ for heads on a coin, and p might be 95 percent for a given brand of fire-cracker to explode properly.

Armed with these definitions, we can now ask for the probability of getting ν successes in n trials. A calculation which we will soon sketch shows that this probability is given by the so-called *binomial distribution*:

$$\begin{aligned} P(\nu \text{ successes in } n \text{ trials}) &= b_{n,p}(\nu) \\ &= \frac{n(n-1)\cdots(n-\nu+1)}{1 \times 2 \times \cdots \times \nu} p^\nu q^{n-\nu}. \end{aligned} \tag{10.2}$$

Here the letter b stands for "binomial"; the subscripts n and p on $b_{n,p}(\nu)$ indicate that the distribution depends on n, the number of trials made, and p, the probability of success in one trial.

The distribution (10.2) is called the binomial distribution because of its close connection with the well-known binomial expansion. Specifically, the fraction in (10.2) is the *binomial coefficient*, often denoted $\binom{n}{\nu}$,

$$\binom{n}{\nu} = \frac{n(n-1)\cdots(n-\nu+1)}{1 \times 2 \times \cdots \times \nu} \tag{10.3}$$

$$= \frac{n!}{\nu!(n-\nu)!}, \tag{10.4}$$

where we have introduced the useful *factorial* notation,

$$n! = 1 \times 2 \times \cdots \times n.$$

The binomial coefficient appears in the binomial expansion

$$\begin{aligned} (p+q)^n &= p^n + np^{n-1}q + \cdots + q^n \\ &= \sum_{\nu=0}^{n} \binom{n}{\nu} p^\nu q^{n-\nu}, \end{aligned} \tag{10.5}$$

which holds for any two numbers p and q and any positive integer n (see Problem 10.4).

With the notation (10.3), we can rewrite the binomial distribution in the more compact form

$$P(\nu \text{ successes in } n \text{ trials}) = b_{n,p}(\nu) = \binom{n}{\nu} p^{\nu} q^{n-\nu}, \tag{10.6}$$

where, as usual, p denotes the probability of success in one trial and $q = 1 - p$.

The derivation of the result (10.6) is similar to that of the example of the dice in (10.1),

$$P(2 \text{ aces in } 3 \text{ throws}) = 3 \times (\tfrac{1}{6})^2 \times (\tfrac{5}{6}). \tag{10.7}$$

In fact, if we set $\nu = 2$, $n = 3$, $p = \frac{1}{6}$, and $q = \frac{5}{6}$ in (10.6), we obtain precisely (10.7), as you should check. Furthermore, the significance of each factor in (10.6) is the same as that of the corresponding factor in (10.7). The factor p^{ν} is the probability of getting all successes in any definite ν trials, and $q^{n-\nu}$ is the probability of failure in the remaining $n - \nu$ trials. The binomial coefficient $\binom{n}{\nu}$ is easily shown to be the number of different orders in which one can get ν successes in n trials. This establishes that the binomial distribution (10.6) is indeed the probability claimed.

Example

Suppose we toss four coins ($n = 4$) and count the number of heads obtained, ν. What is the probability of obtaining the various possible values $\nu = 0, 1, 2, 3, 4$?

Since the probability of getting a head on one toss is $p = \frac{1}{2}$, the required probability is simply the binomial distribution $b_{n,p}(\nu)$, with $n = 4$ and $p = q = \frac{1}{2}$,

$$P(\nu \text{ heads in } 4 \text{ tosses}) = \binom{4}{\nu}\left(\frac{1}{2}\right)^4.$$

These numbers are easily calculated (Problem 10.5) and give the distribution shown in Figure 10.2.

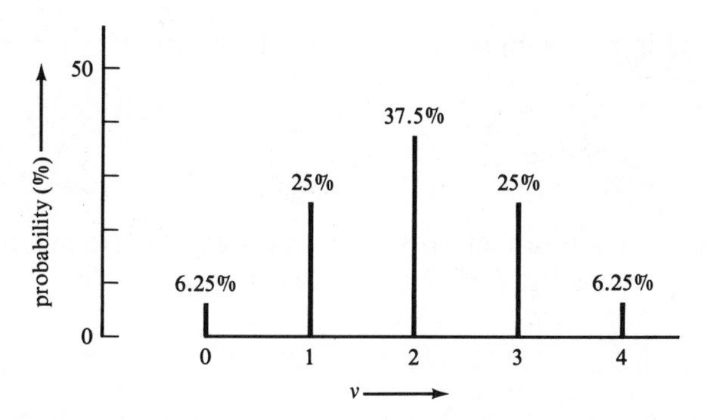

Figure 10.2. The binomial distribution $b_{n,p}(\nu)$ with $n = 4$, $p = \frac{1}{2}$. This gives the probability of getting ν heads when throwing four coins.

We see that the most probable number of heads is $\nu = 2$, as one would expect. Here the probabilities are symmetrical about this most probable value. That is, the probability for three heads is the same as that for one, and the probability for four heads is the same as that for none. As we will see, this symmetry occurs only when $p = \frac{1}{2}$.

10.4. Properties of the Binomial Distribution

The binomial distribution $b_{n,p}(\nu)$ gives the probability of having ν "successes" in n trials, when p is the probability of success in a single trial. If we repeat our whole experiment (consisting of n trials) many times, then it is natural to ask what would be our average number of successes $\bar{\nu}$. This is just

$$\bar{\nu} = \sum_{\nu=0}^{n} \nu b_{n,p}(\nu), \tag{10.8}$$

and is easily evaluated (Problem 10.8) as

$$\bar{\nu} = np. \tag{10.9}$$

That is, if we repeat our series of n trials many times, then the average number of successes will be just the probability of success in one trial (p) times n, as one would expect. One can similarly calculate the standard deviation σ_ν in our number of successes (Problem 10.10). The result is

$$\sigma_\nu = \sqrt{np(1-p)}. \tag{10.10}$$

When $p = \frac{1}{2}$ (as in a coin-tossing experiment) the average number of successes is just $n/2$. Furthermore, it is easy to prove for $p = \frac{1}{2}$ that

$$b_{n,1/2}(\nu) = b_{n,1/2}(n-\nu) \tag{10.11}$$

(see Problem 10.11). That is, the binomial distribution, with $p = \frac{1}{2}$, is symmetrical about the average value $n/2$, as we noticed in Figure 10.2.

In general, when $p \neq \frac{1}{2}$, the binomial distribution $b_{n,p}(\nu)$ is not symmetrical. For example, Figure 10.1 is clearly not symmetrical, the most probable number of successes being $\nu = 0$, and the probability diminishing steadily for $\nu = 1$, 2, and 3. Also, the average number of successes ($\bar{\nu} = 0.5$) is here not the same as the most probable number of successes ($\nu = 0$).

It is interesting to compare the binomial distribution $b_{n,p}(\nu)$ with the more familiar Gauss distribution $f_{X,\sigma}(x)$. Perhaps the biggest difference is that the experiment described by the former has outcomes given by the *discrete*[1] values $\nu = 0, 1, 2, \ldots, n$, whereas those of the latter are given by the *continuous* values of the measured quantity x. The Gauss distribution is a symmetrical peak centered on the average value $x = X$, which means that the average value X is also the most probable value (that for which $f_{X,\sigma}(x)$ is maximum). As we have seen, the binomial distribution is symmetrical only when $p = \frac{1}{2}$, and in general the average value does not coincide with most probable value.

Gaussian Approximation to the Binomial Distribution

For all their differences, there is an important connection between the binomial and Gauss distributions. If we consider the binomial distribution $b_{n,p}(\nu)$ for any fixed value of p, then when n is large $b_{n,p}(\nu)$ is closely approximated by the Gauss distribution $f_{X,\sigma}(\nu)$ with the same mean and same standard deviation; that is,

$$b_{n,p}(\nu) \approx f_{X,\sigma}(\nu) \qquad (n \text{ large}) \tag{10.12}$$

[1] The word *discrete* (not to be confused with discreet) means "detached from one another" and is the opposite of continuous.

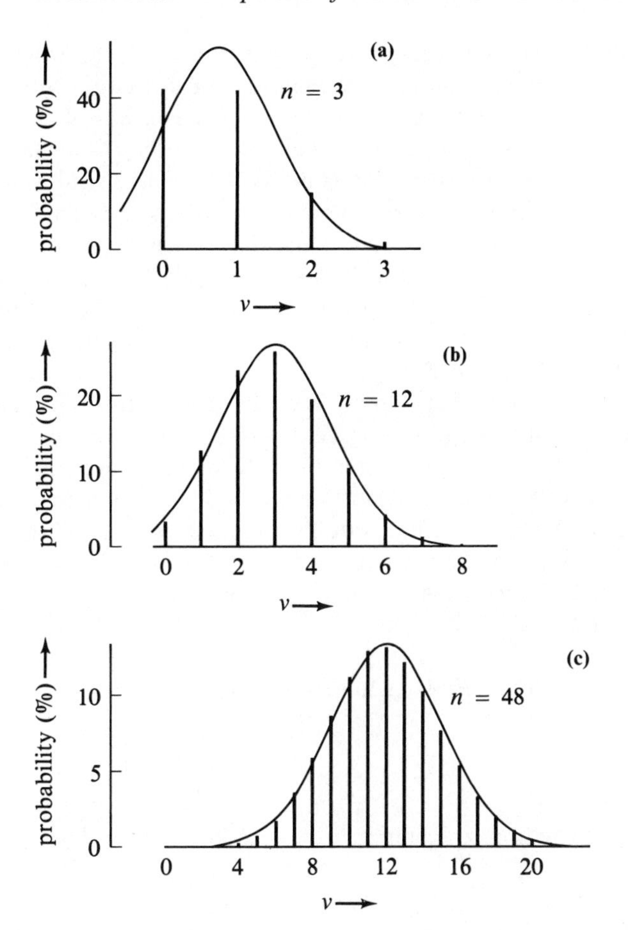

Figure 10.3. The binomial distributions with $p = \frac{1}{4}$ and $n = 3$, 12, and 48. The continuous curve superimposed on each picture is the Gauss function with the same mean and same standard deviation.

with

$$X = np \quad \text{and} \quad \sigma = \sqrt{np(1-p)}. \tag{10.13}$$

We refer to (10.12) as the Gaussian approximation to the binomial distribution. We shall not prove the result here,[2] but its truth is clearly illustrated in Figure 10.3, which shows the binomial distribution for $p = \frac{1}{4}$

[2] For proofs, see Stuart L. Meyer, *Data Analysis for Scientists and Engineers* (John Wiley, 1975), p. 226, or Hugh D. Young, *Statistical Treatment of Experimental Data* (McGraw-Hill, 1962), Appendix C.

and for three successively larger values of n ($n = 3, 12, 48$). Superimposed on each binomial distribution is the Gaussian distribution with the same mean and standard deviation. With just three trials ($n = 3$), the binomial distribution is quite different from the corresponding Gaussian. In particular, the binomial distribution is distinctly asymmetrical, whereas the Gaussian is, of course, perfectly symmetrical about its mean. By the time $n = 12$, the asymmetry of the binomial distribution is much less pronounced, and the two distributions are quite close to one another. When $n = 48$, the difference between the binomial and the corresponding Gauss distribution is so slight that the two are almost indistinguishable on the scale of Figure 10.3(c).

That the binomial distribution can be approximated by the Gauss function when n is large is very useful in practice. Calculation of the binomial function with n greater than 20 or so is extremely tedious, whereas calculation of the Gauss function is always simple whatever the values of X and σ. To illustrate this, suppose we wanted to know the probability of getting 23 heads in 36 tosses of a coin. This probability is given by the binomial distribution $b_{36,1/2}(\nu)$, since the probability of a head in one toss is $p = \frac{1}{2}$. Thus

$$P(\text{23 heads in 36 tosses}) = b_{36,1/2}(23) \tag{10.14}$$

$$= \frac{36!}{23!13!}\left(\frac{1}{2}\right)^{36}, \tag{10.15}$$

which a fairly tedious calculation[3] shows to be

$$P(\text{23 heads}) = 3.36\%.$$

On the other hand, since the mean of the distribution is $np = 18$ and the standard deviation is $\sigma = \sqrt{np(1-p)} = 3$, we can approximate (10.14) by the Gauss function $f_{18,3}(23)$, and a trivial calculation gives

$$P(\text{23 heads}) \approx f_{18,3}(23) = 3.32\%.$$

For almost all purposes this is an excellent approximation.

The usefulness of the Gaussian approximation is even more obvious if we want the probability of several outcomes. For example, the probability

[3] Some hand calculators are preprogrammed to compute $n!$ automatically, and with such a calculator computation of (10.15) is easy. However, in most such calculators $n!$ overflows when $n \geqslant 70$; so for $n \geqslant 70$ this preprogrammed function is no help.

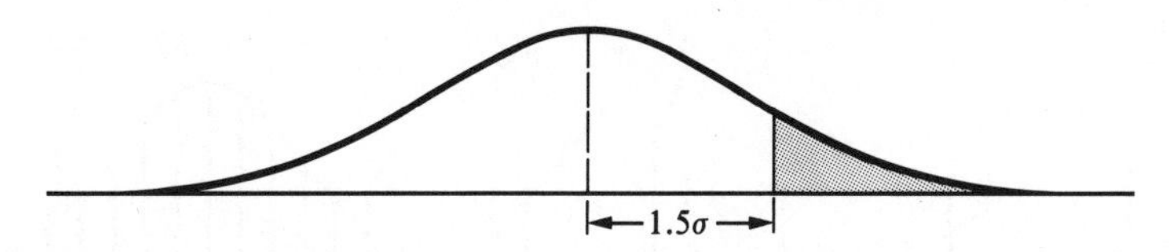

Figure 10.4. The probability of a result more than 1.5σ above the mean is the shaded area under the Gauss curve.

of getting 23 *or more* heads in 36 tosses is

$$P(23 \text{ or more heads}) = P(23 \text{ heads}) + P(24 \text{ heads}) + \cdots + P(36 \text{ heads}),$$

a most tedious sum to calculate directly. However, if we approximate the binomial distribution by the Gaussian, then the probability is easily found. Since the calculation of Gaussian probabilities treats v as a continuous variable, the probability for $v = 23, 24, \ldots$ is best calculated as $P_{\text{Gauss}}(v \geqslant 22.5)$, the probability for any $v \geqslant 22.5$. Now, $v = 22.5$ is 1.5 standard deviations above the mean value, 18. (Remember, $\sigma = 3$; so $4.5 = 1.5\sigma$.) The probability of a result more than 1.5σ above the mean equals the area under the Gauss function shown in Figure 10.4. It is easily calculated with the help of the table in Appendix B, and we find

$$P(23 \text{ or more heads}) \approx P_{\text{Gauss}}(v \geqslant X + 1.5\sigma) = 6.7\%.$$

This compares well with the exact result (to two significant figures) 6.6 percent.

10.5. The Gauss Distribution for Random Errors

In Chapter 5 it was stated that a measurement subject to many small random errors will be normally distributed. We are now in a position to prove this, using a simple model for the kind of measurement concerned.

Let us suppose that we measure a quantity x whose true value is X. We assume that our measurements are subject to negligible systematic error, but that there are n independent sources of random error (effects of parallax, reaction times, and so on). To simplify our discussion, suppose further that all these sources produce random errors of the same fixed size ε. That is, each source of error pushes our result upward by ε or downward by ε, and these two possibilities occur with equal probability,

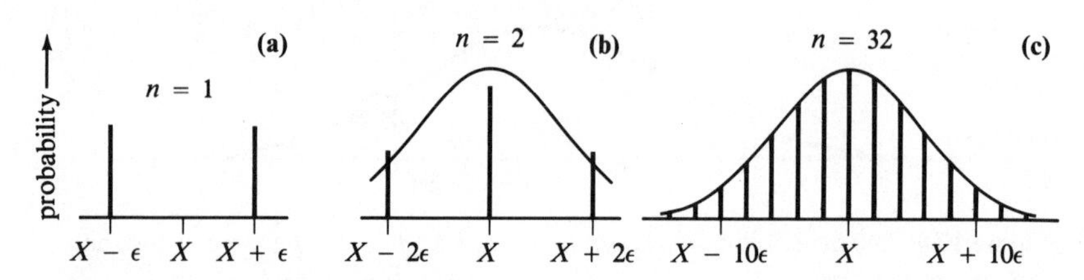

Figure 10.5. Distribution of measurements subject to n random errors of magnitude ε, for $n = 1, 2$, and 32. The continuous curves superimposed on (b) and (c) are Gaussians with the same center and width. (The vertical scales differ in the three pictures.)

$p = \frac{1}{2}$. For example, if the true value is X and there is just one source of error, then our possible answers are $x = X - \varepsilon$ and $x = X + \varepsilon$, both being equally likely. If there are two sources of error, a measurement could yield $x = X - 2\varepsilon$ (if both errors happened to be negative), or $x = X$ (if one was negative and one positive), or $x = X + 2\varepsilon$ (if both happened to be positive). These possibilities are shown in Figures 10.5(a) and (b).

In general, if there are n sources of error, our answer could range from $x = X - n\varepsilon$ to $x = X + n\varepsilon$. In a given measurement, if ν sources happen to give positive errors and $(n - \nu)$ negative errors, then our answer will be

$$\begin{aligned} x &= X + \nu\varepsilon - (n - \nu)\varepsilon \\ &= X + (2\nu - n)\varepsilon. \end{aligned} \tag{10.16}$$

The probability of this occurring is just the binomial probability

$$P(\nu \text{ positive errors}) = b_{n,1/2}(\nu). \tag{10.17}$$

Thus the possible results of our measurement are symmetrically distributed around the true value X, with probabilities given by the binomial function (10.17). This is illustrated in Figure 10.5 for $n = 1, 2$, and 32.

What we now claim is that if the number of sources of error, n, is large and the size of the individual errors, ε, is small, then our measurements are normally distributed. To be more precise, we note that the standard deviation of the binomial distribution is $\sigma_\nu = \sqrt{np(1-p)} = \sqrt{n/4}$. Therefore, according to (10.16) the standard deviation of our measurements of x is $\sigma_x = 2\varepsilon\sigma_\nu = \varepsilon\sqrt{n}$. Accordingly, we let $n \to \infty$ and $\varepsilon \to 0$ in such a way that $\sigma_x = \varepsilon\sqrt{n}$ remains fixed. When we do this, two things happen. First, as discussed in the previous section, the binomial distribution approaches the Gauss distribution with center X and width σ_x. This is clearly visible in Figures 10.5(b) and (c), on which the appropriate Gauss functions have been superimposed. Second, as $\varepsilon \to 0$, the possible results of our measurement get closer together (as is also clear in Figure 10.5), so that the discrete distribution goes over to a continuous distribution, which is precisely the expected Gauss distribution.

10.6. Applications; Testing of Hypotheses

Once we know how the results of an experiment should be distributed, we can ask whether the actual results of the experiment *were* distributed as expected. This kind of test of a distribution is an important technique in the physical sciences, and is perhaps even more important in the biological and social sciences. One important and general test, the χ^2 test, is the subject of Chapter 12. Here we give two examples of a simpler test that can be applied to certain problems involving the binomial distribution.

Testing a New Ski Wax

Suppose that a manufacturer of ski waxes claims to have developed a new wax that greatly reduces the friction between skis and snow. To test this claim, we might take ten pairs of skis and treat one ski from each pair with the wax. We could then hold races between the treated and untreated members of each pair, by letting them slide down a suitable snow-covered incline.

If the treated skis won all ten races, we would obviously have strong evidence that the wax works. Unfortunately, we seldom get such a clear-cut result; and even when we do, we would like some quantitative measure of the strength of the evidence. Thus we have to address two questions. First, how can we quantify the evidence that the wax works (or doesn't work)? Second, where would we draw the line? If the treated skis won nine of the races, would this be conclusive? And what if they won eight races? Or seven?

Precisely these same questions arise in a host of similar statistical tests. If we wished to test the efficacy of a fertilizer, we would organize "races" between treated and untreated plants. To predict which candidate is going to win an election, we would choose a random sample of voters and hold "races" between the candidates with the members of our sample.

To answer our questions, we need to decide more precisely what we should expect from our tests. In the accepted terminology, we must formulate a *statistical hypothesis.* In the example of the ski wax, the simplest hypothesis is the *null hypothesis,* that the new wax actually makes no difference. Subject to this hypothesis we can calculate the probability of the various possible results of our test, and then judge the significance of our particular result.

Suppose we take as our hypothesis that the ski wax makes no difference. In any one race the treated and untreated skis would then be equally likely to win; that is, the probability for a treated ski to win is $p = \frac{1}{2}$. The probability that the treated skis will win ν of the ten races is then the binomial probability:

$$\begin{aligned} P(\nu \text{ wins in 10 races}) &= b_{10,1/2}(\nu) \\ &= \frac{10!}{\nu!(10-\nu)!}\left(\frac{1}{2}\right)^{10}. \end{aligned} \tag{10.18}$$

According to (10.18) the probability that the treated skis would win all ten races is

$$P(10 \text{ wins in 10 races}) = (\tfrac{1}{2})^{10} \approx 0.1\%. \tag{10.19}$$

That is, if our null hypothesis is correct, it is *very* unlikely that the treated skis would win all ten races. Conversely, if the treated skis *did* win all ten races, it is very unlikely the null hypothesis is correct. In fact, the probability (10.19) is so small we could say the evidence in favor of the wax is "highly significant," as we will discuss shortly.

Suppose instead that the treated skis had won eight of the ten races. Here we would calculate the probability of eight *or more* wins:

$$\begin{aligned} P(8 \text{ or more wins in 10 races}) &= P(8 \text{ wins}) + P(9 \text{ wins}) + P(10 \text{ wins}) \\ &\approx 5.5\%. \end{aligned} \tag{10.20}$$

For the treated skis to win eight or more races is still quite unlikely, but not nearly as unlikely as their winning all ten.

To decide what conclusion to draw from the eight wins, we must recognize that there are really just two alternatives: Either

(a) our null hypothesis is correct (the wax makes no difference), but, by chance, an unlikely event has occurred (the treated skis have won eight races);

or

(b) our null hypothesis is false, and the wax does help.

In statistical testing it is traditional to pick some definite probability (e.g., 5 percent) that one will regard as defining the boundary below which an event is unacceptably improbable. If the probability of the actual outcome (eight or more wins, in our case) is below this boundary, then we choose alternative (b), we reject the hypothesis, and we say that the result of our experiment was *significant*.

It is a common practice to call a result significant if its probability is less than 5 percent, and to call it "highly significant" if its probability is less than 1 percent. Since the probability (10.20) is 5.5 percent, we see that eight wins for the waxed skis are just *not* enough to give "significant" evidence that the wax works. On the other hand, we saw that the probability of ten wins is 0.1 percent. Since this is less than 1 percent, we can say that ten wins would constitute "highly significant" evidence that the wax helps.[4]

General Procedure

The methods of the example just described can be applied to any set of n similar but independent tests (or "races"), each of which has the same two possible outcomes, "success" or "failure." One first formulates a hypothesis, here simply an assumed value for the probability p of success in any one test. This assumed value of p determines the expected mean number of successes, $\bar{\nu} = np$, in n trials.[5] If the actual number of successes, ν, in our n trials is close to np, then there is no evidence against the hypothesis. (If the waxed skis win five out of ten races, there is no evidence that the wax makes any difference.) If ν is appreciably larger than np, then we calculate the probability (given our hypothesis) of getting ν or more successes. If this probability is less than our chosen "significance level" (e.g. 5 percent or 1 percent), then we argue that our observed number is unacceptably improbable (if our hypothesis is correct) and

[4] It is perhaps worth emphasizing the great simplicity of the test just described. We could have measured various additional parameters, such as the time taken by each ski, the maximum speed of each ski, and so on. Instead, we simply recorded which ski won each race. Tests that do not use such additional parameters are called *nonparametric* tests. They have the great advantages of simplicity and wide applicability.

[5] As usual, $\bar{\nu} = np$ is the mean number of successes expected if we were to repeat our whole set of n trials many times.

hence that our hypothesis should be rejected. In the same way, if our number of successes ν is appreciably less than np, then we can argue similarly, except that we would calculate the probability of getting ν or less successes.[6]

As we should have expected, this procedure does not provide a simple answer that our hypothesis is certainly true or certainly false. What it does give is a quantitative measure of the reasonableness of our results in the light of the hypothesis; so we are able to choose an objective, if arbitrary, criterion for rejection of the hypothesis. When an experimenter states a conclusion based on this kind of reasoning, it is important that he or she state clearly what criterion was used and what the calculated probability was, so that the reader can judge the reasonableness of the conclusions.

An Opinion Poll

As a second example, consider an election between two candidates, A and B. Suppose candidate A claims that extensive research has established that he is favored by 60 percent of the electorate, and suppose candidate B asks us to check this claim (in the hope, of course, of showing that the number favoring A is significantly less than 60 percent).

Here our statistical hypothesis would be that 60 percent of voters favor A; so the probability that a randomly selected voter will favor A would be $p = 0.6$. Recognizing that we cannot poll every single voter, we carefully select a random sample of 600 and ask their preferences. If 60 percent really favor A, then the expected number in our sample who favor A is $np = 600 \times 0.6 = 360$. If in fact 330 state a preference for A, can we claim to have cast significant doubt on the hypothesis that 60 percent favor A?

To answer this question, we note that (according to the hypothesis) the probability that ν voters will favor A is the binomial probability

$$P(\nu \text{ voters favor } A) = b_{n,p}(\nu) \tag{10.21}$$

with $n = 600$ and $p = 0.6$. Because n is so large, it is an excellent approximation to replace the binomial function by the appropriate Gauss function, with center at $np = 360$ and standard deviation $\sigma_\nu = \sqrt{np(1-p)} = 12$,

$$P(\nu \text{ voters favor } A) \approx f_{360,12}(\nu). \tag{10.22}$$

[6] As we discuss below, in some experiments the relevant probability is the "two-tailed" probability of getting a value of ν that deviates *in either direction* from np by as much as, or more than, the value actually obtained.

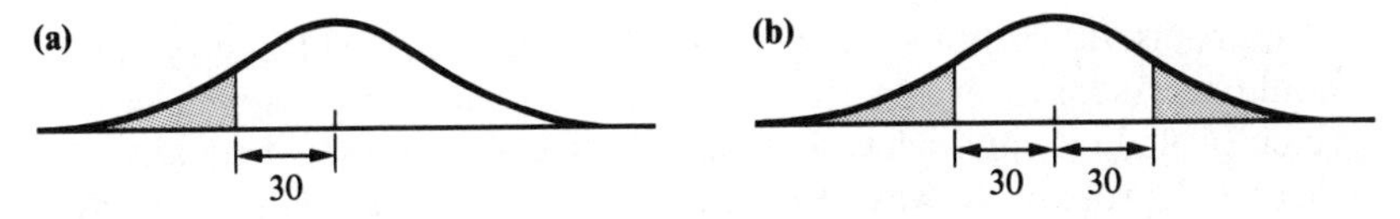

Figure 10.6. (a) The "one-tailed" probability for getting a result 30 or more below the mean. (b) The "two-tailed" probability for getting a result that differs from the mean by 30 or more in either direction. (Not to scale.)

The mean number expected to favor A is 360. Thus the number who actually favored A in our sample (namely 330) is 30 less than expected. Since the standard deviation is 12, our result is 2.5 standard deviations below the supposed mean. The probability of a result this low or lower (according to the table in Appendix B) is 0.6 percent.[7] Thus our result is "highly significant," and at the 1 percent level we can confidently reject the hypothesis that A is favored by 60 percent.

This example illustrates two general features of this kind of test. First, having found that 330 voters favored A (that is, 30 less than expected), we calculated the probability that the number favoring A would be 330 *or less*. At first thought, one might have considered the probability that the number favoring A is precisely $\nu = 330$. However, this probability is extremely small (0.15 percent, in fact), and even the most probable result ($\nu = 360$) has a low probability (3.3 percent). To get a proper measure of how unexpected the result $\nu = 330$ is, we have to include $\nu = 330$ *and* any result that is even further below the mean.

Our result $\nu = 330$ was 30 less than the expected result, 360. The probability of a result that is 30 or more below the mean is sometimes called a "one-tailed probability," since it is the area under one tail of the distribution curve, as in Figure 10.6(a). In some tests the relevant probability is the "two-tailed probability" of getting a result that differs from the expected mean by 30 or more *in either direction*; that is, the probability of getting $\nu \leqslant 330$ *or* $\nu \geqslant 390$, as in Figure 10.6(b). Whether one uses the one-tailed or the two-tailed probability in a statistical test depends on what one considers the interesting alternative to the original hypothesis. Here we were concerned to show that candidate A was favored by *less* than the claimed 60 percent; so the one-tailed probability was appropriate. If we were concerned to show that the number favoring A was *different* from 60 percent (in either direction), then the two-tailed probability would

[7] Strictly speaking, we should have computed the probability for $\nu \leqslant 330.5$ because the Gauss distribution treats ν as a continuous variable. This is 2.46σ below the mean; so the correct probability is actually 0.7 percent, but this small a difference does not affect our conclusion.

be appropriate. In practice, it is usually fairly clear which probability should be used. In any case, the experimenter always needs to state clearly which probability and what significance level were chosen, and what the calculated probability was. With this information the reader can judge the significance of the results for himself or herself.

Problems

Reminder: An asterisk indicates that the problem is discussed, or its answer given, in the Answers section at the back of the book.

10.1 (Section 10.2). Consider the experiment of Section 10.2 in which three dice are thrown. Derive the probabilities for throwing no aces and for throwing one ace. Verify all four of the probabilities shown in Figure 10.1.

***10.2** (Section 10.2).
(a) Compute the probabilities $P(\nu \text{ aces in two throws})$ for all possible ν in a throw of two dice. Plot them in a histogram.
(b) Do the same for a throw of four dice.

10.3 (Section 10.3).
(a) Compute $5!$, $6!$, $25!/23!$
(b) Use the relation $n! = (n+1)!/(n+1)$ to show that $0!$ should be defined as 1.
(c) Prove that the binomial coefficient defined by Equation (10.3) is equal to

$$\binom{n}{\nu} = \frac{n!}{\nu!(n-\nu)!}.$$

***10.4** (Section 10.3). Evaluate the binomial coefficients $\binom{3}{\nu}$ for $\nu = 0, 1, 2, 3$, and $\binom{4}{\nu}$ for $\nu = 0, \ldots, 4$. Hence write down the binomial expansion, (10.5), of $(p+q)^n$ for $n = 3$ and 4.

10.5 (Section 10.3).
(a) Compute and plot a histogram of the binomial distribution function $b_{n,p}(\nu)$ for $n = 4$, $p = \frac{1}{2}$, and all possible ν.
(b) Repeat part (a) for $n = 4$ and $p = \frac{1}{5}$.

***10.6** (Section 10.3). A hospital admits four patients suffering from a disease, the mortality rate of which is 80 percent. Use the results of

Problem 10.5(b) to find the probabilities of the following outcomes:

(a) none of the patients survive;
(b) exactly one survives;
(c) two or more survive.

***10.7** (Section 10.3). Find the probabilities for getting ν aces in a throw of five dice for $\nu = 0, 1, \ldots, 5$.

10.8 (Section 10.4). Prove that the mean number of successes

$$\bar{\nu} = \sum_{\nu=0}^{n} \nu b_{n,p}(\nu)$$

for the binomial distribution is just np.

There are many ways to do this, of which one of the best is this: Write down the binomial expansion (10.5) for $(p+q)^n$. Since this is true for any p and q, you can differentiate with respect to p. If you now set $p + q = 1$, and multiply through by p, you will have the desired result.

***10.9** (Section 10.4). The standard deviation for any distribution $f(\nu)$ is defined by

$$\sigma_\nu^{\,2} = \overline{(\nu - \bar{\nu})^2}.$$

Prove that this is the same as $\overline{\nu^2} - (\bar{\nu})^2$.

***10.10** (Section 10.4). Use the result of Problem 10.9 to prove for the binomial distribution $b_{n,p}(\nu)$ that

$$\sigma_\nu^{\,2} = np(1-p).$$

(Use the same trick as in Problem 10.8, but differentiate with respect to p twice.)

10.11 (Section 10.4). Prove that for $p = \frac{1}{2}$ the binomial distribution satisfies

$$b_{n,1/2}(\nu) = b_{n,1/2}(n - \nu);$$

that is, the distribution is symmetrical about $\nu = n/2$.

10.12 (Section 10.4). The Gaussian approximation (10.12) to the binomial distribution is excellent for n large, and surprisingly good for n small (especially if p is close to $\frac{1}{2}$). To illustrate this, calculate $b_{4,1/2}(\nu)$ (for $\nu = 0, 1, \ldots, 4$), both exactly and using the Gaussian approximation. Compare your results.

***10.13** (Section 10.4). Use the Gaussian approximation to find the probability of getting exactly 15 heads if you throw a coin 25 times. Calculate the same probability exactly, and compare answers.

***10.14** (Section 10.4). Use the Gaussian approximation to find the probability of getting 18 or more heads in 25 tosses of a coin. (In using the Gauss probabilities, you must find the probability for $\nu \geqslant 17.5$.) Compare with the exact answer, which is 2.16 percent.

10.15 (Section 10.6). In the test of a ski wax described in Section 10.6, suppose the waxed skis had won nine out of the ten races. Assuming that the wax makes no difference, calculate the probability of nine or more wins. Do nine wins give "significant" evidence that the wax is effective (5 percent level)? Is the evidence "highly significant" (1 percent level)?

***10.16** (Section 10.6). To test a new fertilizer, a gardener selects 14 pairs of similar plants, and treats one plant from each pair with the fertilizer. After two months, 12 of the treated plants are healthier than their untreated partners (and the remaining two are less healthy). If in fact the fertilizer made no difference, what would be the probability that pure chance led to 12 or more successes? Do the 12 successes give significant evidence that the fertilizer helps (5 percent level)? Is the evidence "highly significant" (1 percent level)?

10.17 (Section 10.6). It is known that 25 percent of a certain kind of seed normally germinates. To test a new "germination stimulant," 100 of these seeds are planted and treated with the stimulant. If 32 of them germinate, can one conclude (at the 5 percent level of significance) that the stimulant helps?

***10.18** (Section 10.6). In a certain school, 420 of the 600 students pass a standardized mathematics test, for which the national passing rate is 60 percent. If the students of the school had no special aptitude for the test, how many would you expect to pass, and what is the probability that 420 or more would pass? Can the school claim that its students are significantly well-prepared for the test?

CHAPTER 11

The Poisson Distribution

In this chapter we study our third example of a limiting distribution function, the Poisson distribution. It describes the results of experiments where one counts events that occur at random, but at a definite average rate. It is especially important in atomic and nuclear physics, where one counts the disintegrations of unstable atoms and nuclei.

11.1. Definition of the Poisson Distribution

As an example of the Poisson distribution, suppose we are given a sample of radioactive material. Using a Geiger counter, we can count the number ν of electrons ejected by radioactive decays in a period of one minute. If the counter is reliable, then there will be no uncertainty in our value of ν. Nevertheless, if we repeat the experiment we will almost certainly get a different answer for ν. This variation in the number ν does not reflect uncertainties in our counting, but is rather an intrinsic property of the radioactive-decay process.

Each radioactive nucleus has a definite probability for decaying in any one-minute interval. If we knew this probability and the number of nuclei in our sample, then we could calculate the *expected average number* of decays in a minute. Nevertheless, each nucleus decays at a random time, and in any one minute the number of decays may be different from the expected average number.

Obviously the question we should be asking is this: If we repeat our experiment many times (replenishing our sample if it becomes significantly depleted), what distribution should we expect for the number of decays ν observed in one-minute intervals? If you have studied Chapter 10, you will recognize that the required distribution is the binomial distribution. If there are n nuclei, and the probability that any one nucleus decays is p, then the probability of ν decays is just the probability of ν "successes" in

n "trials", or $b_{n,p}(\nu)$. However, in the kind of experiment we are now discussing, there is an important simplification. The number of "trials" (i.e., nuclei) is enormous ($n \sim 10^{20}$, perhaps), and the probability of "success" (i.e., decay) for any one nucleus is tiny (often as small as $p \sim 10^{-20}$ or so). Under these conditions (n large and p small), it can be shown that the binomial distribution is indistinguishable from a simpler function, called the Poisson distribution. Specifically, one can prove that

$$P(\nu \text{ counts in any definite interval}) = p_\mu(\nu), \tag{11.1}$$

where the *Poisson distribution*, $p_\mu(\nu)$, is given by

$$p_\mu(\nu) = e^{-\mu} \frac{\mu^\nu}{\nu!}. \tag{11.2}$$

In this definition, μ is a positive parameter ($\mu > 0$) which, we will soon see, is just the expected mean number of counts in the time interval concerned; and $\nu!$ denotes the usual factorial function (and $0! = 1$).

We will not derive the Poisson distribution (11.2) here, but will simply assert that it is the appropriate distribution for the kind of experiment being discussed.[1] To establish the significance of the parameter μ in (11.2), we have only to calculate the average number of counts $\bar{\nu}$ expected if we repeat our counting experiment many times. This average is

$$\bar{\nu} = \sum_{\nu=0}^{\infty} \nu p_\mu(\nu) = \sum_{\nu=0}^{\infty} \nu e^{-\mu} \frac{\mu^\nu}{\nu!}. \tag{11.3}$$

The first term of this sum can be dropped (since it is zero), and $\nu/\nu!$ can be replaced by $1/(\nu-1)!$. If we remove a common factor of $\mu e^{-\mu}$, this gives

$$\bar{\nu} = \mu e^{-\mu} \sum_{\nu=1}^{\infty} \frac{\mu^{\nu-1}}{(\nu-1)!}. \tag{11.4}$$

The infinite sum that remains is

$$1 + \mu + \frac{\mu^2}{2!} + \frac{\mu^3}{3!} + \cdots = e^\mu, \tag{11.5}$$

[1] For derivations, see, for example, Hugh D. Young, *Statistical Treatment of Experimental Data* (McGraw-Hill, 1962), Section 8, or Stuart L. Meyer, *Data Analysis for Scientists and Engineers* (John Wiley, 1975), p. 207.

which is just the exponential function e^{μ} (as indicated). Thus the exponential $e^{-\mu}$ in (11.4) is exactly canceled by the sum, and we are left with the simple conclusion that

$$\bar{\nu} = \mu. \tag{11.6}$$

That is, the parameter μ that characterizes the Poisson distribution $p_{\mu}(\nu)$ is just the *average number of counts expected if we repeat the counting experiment many times.*

11.2. Properties of the Poisson Distribution

In Figure 11.1 are plotted the Poisson distributions for the cases $\mu = 0.8$ and 3. In Figure 11.1(a), with $\mu = 0.8$, we see that the most probable counts are $\nu = 0$ or 1 (with $\nu = 0$ slightly more probable), but that there is an appreciable probability of getting $\nu = 2$ or 3. In Figure 11.1(b), with $\mu = 3$, the most probable counts are 2 and 3, with an appreciable probability of counts ranging from $\nu = 0$ all the way to $\nu = 7$. In both pictures the distribution is visibly asymmetrical.

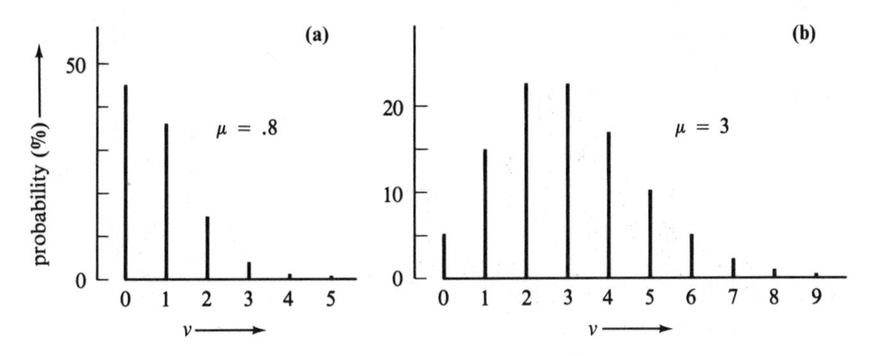

Figure 11.1. Poisson distributions with average counts $\mu = 0.8$ and 3.

If we consider an experiment with a larger average count, e.g., $\mu = 9$, as shown in Figure 11.2, then the distribution is more nearly symmetrical about the mean. In fact, it can be proved that as $\mu \to \infty$, the Poisson distribution becomes steadily more symmetrical, and approaches the Gauss distribution with the same mean and standard deviation.[2] In

[2] See Stuart L. Meyer, *op. cit.*, p. 227.

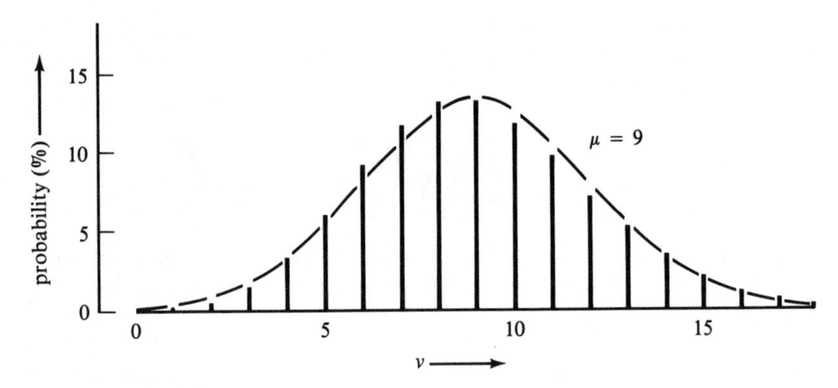

Figure 11.2. The Poisson distribution with $\mu = 9$. The broken curve is the Gauss distribution with the same center and standard deviation.

Figure 11.2, the broken curve is the Gauss function centered on 9 and with the same standard deviation. It can be seen that even when μ is only 9, the Poisson distribution is remarkably close to the appropriate Gauss function, the slight discrepancy reflecting the remaining asymmetry in the Poisson function. As we will shortly discuss, it is very convenient in practice that for large μ one can approximate the Poisson distribution with an appropriate Gaussian.

Another interesting property of the Poisson distribution emerges if we calculate its standard deviation σ_ν. As we saw in Chapter 4, $\sigma_\nu{}^2$ is the average of the squared deviations $(\nu - \bar{\nu})^2$. Thus

$$\sigma_\nu{}^2 = \overline{(\nu - \bar{\nu})^2}$$

or (using the result of Problem 10.9)

$$\sigma_\nu{}^2 = \overline{\nu^2} - (\bar{\nu})^2. \tag{11.7}$$

We have already calculated $\bar{\nu}$ as μ, and a similar calculation gives $\overline{\nu^2} = \mu^2 + \mu$ (see Problem 11.6). Thus $\sigma_\nu{}^2 = \mu$ or

$$\sigma_\nu = \sqrt{\mu}. \tag{11.8}$$

That is, the Poisson distribution with mean count μ has standard deviation $\sqrt{\mu}$.

The result (11.8) is extremely useful in practice. If we carry out a counting experiment once and get the answer ν, then it is easily seen (using the principle of maximum likelihood, as in Problem 11.9) that our best estimate for the expected mean count μ is $\mu_{\text{best}} = \nu$. From (11.8) it immediately follows that our best estimate for the standard deviation is just $\sqrt{\nu}$. In other words, if we make one measurement of a number of events in some time interval and get the answer ν, then our final answer for the expected mean count for that time interval is

$$\nu \pm \sqrt{\nu}. \tag{11.9}$$

This is the result quoted without proof in Equation (3.2). If we were to count for a longer time, we would get a larger value of ν. According to (11.9) this would mean a larger uncertainty $\sqrt{\nu}$. However, the fractional uncertainty, which is

$$\text{fractional uncertainty} = \frac{\sqrt{\nu}}{\nu} = \frac{1}{\sqrt{\nu}},$$

would *decrease*, if we counted for a longer time.

It is interesting to compare the Poisson and Gauss distributions. First, where the Gauss distribution $f_{X,\sigma}(x)$ is *continuous*, since x is a continuous variable, the Poisson distribution $p_\mu(\nu)$ is *discrete* (like the binomial) since $\nu = 0, 1, 2, \ldots$. Second, the Gauss distribution $f_{X,\sigma}(x)$ is defined by *two* parameters, X, the mean, and σ, the width; but the Poisson distribution $p_\mu(\nu)$ is defined by a single parameter, μ, because, as we have just seen, the width σ_ν of the Poisson distribution is automatically determined by the mean μ (namely, $\sigma_\nu = \sqrt{\mu}$). Finally, if we consider Poisson distributions for which the mean count μ is large, then the discrete nature of ν becomes less important, and, as discussed in connection with Figure 11.2, the Poisson distribution (like the binomial) is well-approximated by the Gauss function $f_{X,\sigma}(\nu)$ with the same mean and width. That is,

$$p_\mu(\nu) \approx f_{X,\sigma}(\nu) \qquad (\mu \text{ large}) \tag{11.10}$$

where

$$X = \mu \quad \text{and} \quad \sigma = \sqrt{\mu}.$$

The approximation (11.10) is called the Gaussian approximation to the Poisson distribution. It is analogous to the corresponding approximation for the binomial distribution (discussed in Section 10.4) and is useful under the same conditions, namely, when the parameters involved are large. To illustrate this, suppose we wished to calculate the Poisson distribution with $\mu = 64$. The probability of 72 counts, for example, is

$$P(72 \text{ counts}) = p_{64}(72) = e^{-64}\frac{(64)^{72}}{72!}, \tag{11.11}$$

which a tedious calculation gives as

$$P(72 \text{ counts}) = 2.9\%.$$

However, according to (11.10), the probability (11.11) is well-approximated as

$$P(72 \text{ counts}) \approx f_{64,8}(72),$$

which is easily evaluated to give

$$P(72 \text{ counts}) \approx 3.0\%.$$

If we wanted to calculate directly the probability of 72 *or more* counts in the same experiment, then an outstandingly tedious calculation would give

$$\begin{aligned} P(\nu \geqslant 72) &= p_{64}(72) + p_{64}(73) + \cdots \\ &= 17.3\%. \end{aligned}$$

If we use the approximation (11.10), then we have only to calculate the probability of getting $\nu \geqslant 71.5$ (since the Gauss distribution treats ν as continuous). Since 71.5 is 7.5, or 0.94σ, above the mean, the required probability can be found immediately from the table in Appendix B as

$$\begin{aligned} P(\nu \geqslant 72) \approx P_{\text{Gauss}}(\nu \geqslant 71.5) &= P_{\text{Gauss}}(\nu \geqslant X + .94\sigma) \\ &= 17.4\%, \end{aligned}$$

by almost any standard an excellent approximation.

11.3. Examples

As we have already emphasized, the Poisson distribution describes the distribution of results in an experiment where one counts events that occur at random, but at a definite, expected average rate. In an introductory physics laboratory, the two commonest examples are counting the disintegrations of radioactive nuclei and counting the arrival of cosmic ray particles.

Another very important example is an experiment to study an expected limiting distribution, like the Gauss or binomial distributions, or the Poisson distribution itself. Any limiting distribution tells us how many events of any particular type are expected when an experiment is repeated several times. (For example, the Gaussian $f_{X,\sigma}(x)$ tells us how many measurements of x are expected to fall in any interval from $x = a$ to $x = b$.) In practice, the observed number is seldom exactly the expected number. Instead it fluctuates in accordance with the Poisson distribution. In particular, if the expected number of events of some type is n, then the observed number can be expected to differ from n by a number of order $\sqrt{n}$.

In many situations it is reasonable to expect numbers to be distributed approximately according to the Poisson distribution. The number of eggs laid in an hour on a poultry farm and the number of births in a day at a hospital would both be expected to follow the Poisson distribution at least approximately (though they would probably show some seasonal variations as well). To test this assumption, you would need to record the number concerned many times over. Plotting the resulting distribution, you could compare it with the Poisson distribution and get a good idea of how well it fitted. For a more quantitative test, you would use the χ^2 test described in Chapter 12.

Cosmic Ray Counting

As a concrete example of the Poisson distribution, let us consider an experiment with cosmic rays. These "rays" are actually charged particles, such as protons and α particles, that enter the Earth's atmosphere from space. Some of them travel all the way to ground level and can be detected (with a Geiger counter, for example) in the laboratory. In the following problem, we exploit the fact that the number of cosmic rays hitting any given area in a given time should follow the Poisson distribution.

Student A asserts that he has measured the number of cosmic rays hitting a Geiger counter in one minute. He claims to have made the measurement repeatedly and carefully, and to have found that, on average, nine particles hit the counter per minute, with "negligible" uncertainty. To check this claim, Student B counts how many particles arrive in one minute and gets the answer 12. Does this cast serious doubt on *A*'s claim that the expected rate is nine?

To make a more careful check, Student *C* counts how many particles arrive in ten minutes. From *A*'s claim she expects to get 90, but actually gets 120. Does this cast significant doubt on *A*'s claim?

Let us consider *B*'s result first. If *A* is right, the expected mean count is 9. Since the distribution of counts should be the Poisson distribution, the standard deviation is $\sqrt{9} = 3$. Student *B*'s result of 12 is therefore only one standard deviation away from the mean of 9. This is certainly not far enough away to contradict *A*'s claim. More specifically, knowing that the probability of any answer ν is supposed to be $p_9(\nu)$, we can calculate the total probability for getting an answer that differs from 9 by 3 or more. This turns out to be 40 percent (see Problem 11.11). Obviously *B*'s result is not at all surprising, and *A* has no reason to worry.

Student *C*'s result is quite a different matter. If *A* is right, then *C* should expect to get 90 counts in 10 minutes. Since the distribution should be Poisson, the standard deviation should be $\sqrt{90} = 9.5$. Thus *C*'s result of 120 is more than *three standard deviations* away from *A*'s prediction of 90. With these large numbers, the Poisson distribution is indistinguishable from the Gauss function, and we can immediately find from the table in Appendix A that the probability of a count more than three standard deviations from the mean is 0.3 percent. That is, if *A* is right, it is extremely improbable that *C* would have observed 120 counts. Turning this around, we can say it is almost certain that something has gone wrong. Perhaps *A* was just not as careful as he claimed. Perhaps the counter was malfunctioning for *A* or *C*, introducing systematic errors into one of the results. Or perhaps *A* made his measurements at a time when the flux of cosmic rays was truly less than normal.

Problems

Reminder: An asterisk indicates that the problem is discussed, or its answer given, in the Answers section at the back of the book.

***11.1** (Section 11.1).

(a) Compute the Poisson distribution $p_\mu(\nu)$ for $\mu = 0.5$ and $\nu = 0, 1, \ldots, 6$, and plot a bar histogram of $p_\mu(\nu)$ against ν.

(b) Repeat part (a) for $\mu = 1$.

(c) Repeat part (a) for $\mu = 2$.

***11.2** (Section 11.1).

(a) The Poisson distribution, like all distribution functions, must satisfy a "normalization condition,"

$$\sum_{\nu=0}^{\infty} p_\mu(\nu) = 1. \tag{11.12}$$

This condition asserts that the total probability of observing *all* possible values of ν must be 1. Prove it. [Remember the infinite series (11.5) for e^μ.]

(b) Differentiate (11.12) with respect to μ, and then multiply by μ to give an alternative proof that $\bar{\nu} = \mu$ as in Equation (11.6).

11.3 (Section 11.1). In the course of 28 days, a poultry farmer finds that between 10 and 10:30 A.M. his brood of hens lays an average of 2.5 eggs. Assuming that the number of eggs laid follows a Poisson distribution with $\mu = 2.5$, on about how many days would you suppose there were no eggs laid between 10 and 10:30? On how many would you suppose there were 2 *or less*? 3 or more?

***11.4** (Section 11.1). A certain radioactive sample contains 1.5×10^{20} nuclei, each of which has a probability $p = 10^{-20}$ of decaying in any given minute.

(a) What is the expected average number, μ, of decays from the sample in one minute?

(b) Compute the probability $p_\mu(\nu)$ of observing ν decays in a minute for $\nu = 0, 1, 2, 3$.

(c) What is the probability of observing four or more decays in one minute?

***11.5** (Section 11.1). A certain radioactive sample is expected to undergo three decays per minute. A student observes the number ν of decays in 100 separate one-minute intervals with the results shown in Table 11.1.

Table 11.1.

number of decays ν	0	1	2	3	4	5	6	7	8	9
times observed	5	19	23	21	14	12	3	2	1	0

(a) Make a histogram of these results, plotting f_ν (the fraction of times the result ν was found) against ν.

(b) On the same plot show the expected distribution $p_3(\nu)$. Do the data seem to fit the expected distribution?

11.6 (Section 11.2).
(a) Prove that the average $\overline{\nu^2}$ for the Poisson distribution $p_\mu(\nu)$ is $\overline{\nu^2} = \mu^2 + \mu$. [The easiest way to do this is probably to differentiate the identity (11.12) twice with respect to μ.]
(b) Hence prove that the standard deviation of ν is $\sigma_\nu = \sqrt{\mu}$. [Use the identity (11.7).]

***11.7** (Section 11.2). Calculate the mean $\bar{\nu}$ and standard deviation σ_ν of the data in Problem 11.5. Compare your answers with the expected values 3 and $\sqrt{3}$.

11.8 (Section 11.2). It is known that the average rate of nuclear disintegrations from a certain sample is roughly 20 per minute. If you wanted to measure this rate within 4 percent, for about how long would you plan to count?

***11.9** (Section 11.2).
(a) Suppose we count the number of cosmic rays hitting a counter in one minute and observe the answer ν_o. Assuming that the number follows the Poisson distribution $p_\mu(\nu)$, where μ is the unknown expected average count, write down the probability of obtaining the observed number ν_o. Use the principle of maximum likelihood to prove that the best estimate for μ is

$$\mu_{\text{best}} = \nu_o.$$

(Remember that the best estimate for μ is that value for which the probability of obtaining the observed ν_o is greatest.)
(b) Suppose we make N separate determinations $\nu_1, \ldots, \nu_N$; argue similarly that here μ_{best} is the average

$$\mu_{\text{best}} = \frac{1}{N} \sum_{i=1}^{N} \nu_i.$$

***11.10** (Section 11.2). The expected mean count in a certain counting experiment is $\mu = 16$.
(a) Use the Gaussian approximation (11.10) to estimate the probability of obtaining ten counts. Compare this with the exact result $p_{16}(10)$.
(b) Use the Gaussian approximation to estimate the probability of getting ten counts *or less*. [Remember to compute $P_{\text{Gauss}}(\nu \leqslant 10.5)$ to allow for the fact that the Gauss distribution treats ν as a continuous variable. The needed probability can be calculated from the table in Appendix B.] Calculate the exact answer and compare.

Note how even with μ as small as 16, the Gaussian approximation gives quite good answers and—at least in part (b)—is significantly less trouble than an exact calculation.

***11.11** (Section 11.3).

(a) Calculate the probabilities $p_9(\nu)$ of obtaining ν counts with $\nu =$ 7, 8, 9, 10, and 11 in an experiment where the expected average count is 9.

(b) Hence find the total probability of obtaining a count which differs from the mean 9 by 3 or more. Would a count of 12 lead you to suspect that the expected mean is not really 9?

11.12 (Section 11.3). A student uses a Geiger counter to measure the activity of a radioactive source. She places the source near the counter, which registers a total of 1600 counts in ten minutes. When she takes the source away, she finds that the counter continues to count, though at a reduced rate. She interprets this continued counting as the result of background radiation, such as cosmic rays and radioactive pollution of the laboratory. To find the background rate she leaves the Geiger counter on for another ten minutes and gets a further 400 counts.

(a) What are the uncertainties in her two answers for the numbers of counts in ten minutes?

(b) Calculate her two counting rates, in counts per minute, and their uncertainties.

(c) Compute the difference of the two answers in part (b) to give her final rate for the source alone and its uncertainty.

CHAPTER 12

The χ^2 Test for a Distribution

We now have had much experience with limiting distributions. These are the functions that describe the expected distribution of results if one repeats an experiment many times. There are many different limiting distributions, corresponding to the many different kinds of experiments that are possible. Perhaps the three most important in physical science are the three which we have already discussed: the Gauss (or normal) function, the binomial distribution, and the Poisson distribution.

In this final chapter we discuss how to decide whether the results of an actual experiment are governed by the expected limiting distribution. Specifically, let us suppose that we perform some experiment for which we believe we know the expected distribution of results. Suppose further that we repeat the experiment several times and record our observations. The question that we now address is this: How can we decide whether our observed distribution is consistent with the expected theoretical distribution? We shall see that this question can be answered using a simple procedure called the *chi-squared*, or χ^2, *test*.

12.1. Introduction to χ^2

Let us begin with a concrete example. Suppose we make 40 measurements $x_1, \ldots, x_{40}$ of the range x of a projectile fired from a certain gun, and get the results shown in Table 12.1. Suppose also we have reason to believe

Table 12.1. Measured values of x (in cm).

731,	772,	771,	681,	722,	688,	653,	757,	733,	742
739,	780,	709,	676,	760,	748,	672,	687,	766,	645
678,	748,	689,	810,	805,	778,	764,	753,	709,	675
698,	770,	754,	830,	725,	710,	738,	638,	787,	712

these measurements are governed by a Gauss distribution $f_{X,\sigma}(x)$, as is certainly very natural. In this type of experiment one usually does not know in advance either the center X or the width σ of the expected distribution. Our first step therefore is to use our 40 measurements to compute best estimates for these quantities:

$$\text{(best estimate for } X) = \bar{x} = \sum_{i=1}^{40} x_i \Big/ 40$$

$$= 730.1 \text{ cm} \qquad (12.1)$$

and

$$\text{(best estimate for } \sigma) = \sqrt{\frac{\sum(x_i - \bar{x})^2}{39}}$$

$$= 46.8 \text{ cm.} \qquad (12.2)$$

Now we can ask whether the actual distribution of our results $x_1, \ldots, x_{40}$ is consistent with our hypothesis that our measurements were governed by the Gauss distribution $f_{X,\sigma}(x)$ with X and σ as estimated. To answer this, we must compute how we would expect our 40 results to be distributed if the hypothesis is true, and compare this expected distribution with our actual observed distribution. The first difficulty is that x is a continuous variable; so we cannot speak of the expected number of measurements equal to any one value of x. Rather, we must discuss the expected number in some interval $a < x < b$. That is, we must divide the range of possible values into *bins*. With 40 measurements, we might choose bin boundaries at $X - \sigma$, X, and $X + \sigma$, giving four bins as in Table 12.2.

We will discuss later the criteria for choosing bins. In particular, they must be chosen so that all bins contain several measured values x_i. In

Table 12.2. A possible choice of bins for the data of Table 12.1. The last line shows the number of data that fell in each bin.

Bin number, k	1	2	3	4
Values of x in bin	$x < X - \sigma$ or $x < 683.3$	$X - \sigma < x < X$ or $683.3 < x < 730.1$	$X < x < X + \sigma$ or $730.1 < x < 776.9$	$X + \sigma < x$ or $776.9 < x$
Observations O_k in bin	8	10	16	6

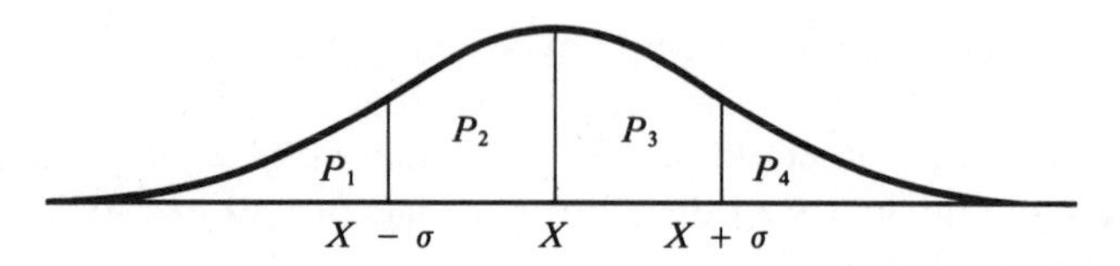

Figure 12.1. The probabilities $P_1, \ldots, P_4$ that a measurement fall in each of the four bins $k = 1, \ldots, 4$ are the four areas shown under the Gauss function.

Table 12.3. Expected numbers E_k and observed numbers O_k for the 40 measurements of Table 12.1.

Bin k	1	2	3	4
Probability P_k	16%	34%	34%	16%
Expected number $E_k = NP_k$	6.4	13.6	13.6	6.4
Observed number O_k	8	10	16	6

general we will denote the number of bins by n; here, for the example with four bins, $n = 4$.

Having divided the range of possible measured values into bins, we can now formulate our question more precisely. First we can count the number of measurements that fall in each bin k.[1] We denote this number by O_k (where O stands for "observed number"). For the data of our example, the observed numbers O_1, O_2, O_3, O_4 are shown in the bottom line of Table 12.2. Next, assuming that our measurements are distributed normally (with X and σ as estimated), we can calculate the *expected* number E_k of measurements in each bin k. We must then decide how well the observed numbers O_k compare with the expected numbers E_k.

The calculation of the expected numbers E_k is quite straightforward. The *probability* that any one measurement fall in an interval $a < x < b$ is just the area under the Gauss function between $x = a$ and $x = b$. In this example, the probabilities P_1, P_2, P_3, P_4 for a measurement to fall in each of our four bins are the four areas indicated in Figure 12.1. The two equal areas P_2 and P_3 together represent the well-known 68 percent; so the probability for falling in one of the two central bins is 34 percent; that is, $P_2 = P_3 = 0.34$. The outside two areas comprise the remaining 32 percent; thus $P_1 = P_4 = 0.16$. To find the expected numbers E_k, we simply multiply these probabilities by the total number of measurements, $N = 40$. Therefore, our expected numbers are as shown in Table 12.3. That the numbers E_k are not integers serves to remind us that the "expected

[1] If a measurement falls exactly on the boundary between two bins, you can assign half a measurement to each bin.

number" is not what we actually expect in any one experiment; it is rather the expected average number when we repeat our whole series of measurements many times.

Our problem now is to decide how well the expected numbers E_k do represent the corresponding observed numbers O_k (on the bottom line of Table 12.3). We would obviously not expect *perfect* agreement between E_k and O_k after any finite number of measurements. On the other hand, if our hypothesis that our measurements are normally distributed is correct, then we would expect that, in some sense, the deviations

$$O_k - E_k \tag{12.3}$$

would be *small*. Conversely, if the deviations $O_k - E_k$ prove to be *large*, then we would suspect that our hypothesis is incorrect.

To make precise the statements that the deviation $O_k - E_k$ is "small" or "large," we must decide how large we would *expect* $O_k - E_k$ to be if the measurements really are normally distributed. Fortunately, this is easily done. If we imagine repeating our whole series of 40 measurements many times, then the number O_k of measurements in any one bin k can be regarded as the result of a counting experiment of the type described in Chapter 11. Our many different answers for O_k should have an average value of E_k and would be expected to fluctuate around E_k with a standard deviation of order $\sqrt{E_k}$. Thus the two numbers to be compared are the deviation $O_k - E_k$ and the expected size of its fluctuations $\sqrt{E_k}$.

These considerations lead us to consider the ratio

$$\frac{O_k - E_k}{\sqrt{E_k}}. \tag{12.4}$$

For some bins k, this ratio will be positive, and for some negative; for a few k, it may be appreciably larger than one, but for most it should be of order one, or smaller. To test our hypothesis (that the measurements are normally distributed), it is natural to square the number (12.4) for each k and then sum over all bins $k = 1, \ldots, n$ (here $n = 4$). This procedure defines a number called *chi squared*,

$$\chi^2 = \sum_{k=1}^{n} \frac{(O_k - E_k)^2}{E_k}. \tag{12.5}$$

It should be clear that this number χ^2 is a reasonable indicator of the agreement between the observed and expected distributions. If $\chi^2 = 0$, the agreement is perfect; that is, $O_k = E_k$ for all bins k, something that is most unlikely to occur. In general, the individual terms in the sum (12.5) are expected to be of order 1, and there are n terms in the sum. Thus if

$$\chi^2 \lesssim n$$

(χ^2 of order n or less), the observed and expected distributions agree about as well as could be expected. In other words, if $\chi^2 \lesssim n$ we have no reason to doubt that our measurements were distributed as expected. On the other hand, if

$$\chi^2 \gg n$$

(χ^2 significantly greater than the number of bins), the observed and expected numbers differ significantly, and there is good reason to suspect that our measurements were not governed by the expected distribution.

In our example, the numbers observed and expected in the four bins are shown again in Table 12.4, and a simple calculation using them gives

$$\begin{aligned}\chi^2 &= \sum_{k=1}^{4} \frac{(O_k - E_k)^2}{E_k} \\ &= \frac{(1.6)^2}{6.4} + \frac{(-3.6)^2}{13.6} + \frac{(2.4)^2}{13.6} + \frac{(-0.4)^2}{6.4} \\ &= 1.80. \end{aligned} \tag{12.6}$$

Since the value of 1.80 for χ^2 is less than the number of terms in the sum (namely, 4), we have no reason to doubt our hypothesis that our measurements were distributed normally.

Table 12.4. Calculation of χ^2 for the data of Table 12.1.

Bin number k	1	2	3	4
Bin	$x < X - \sigma$	$X - \sigma < x < X$	$X < x < X + \sigma$	$X + \sigma < x$
Observed number O_k	8	10	16	6
Expected number E_k	6.4	13.6	13.6	6.4
$O_k - E_k$	1.6	-3.6	2.4	-0.4

12.2. *General Definition of χ^2*

Our discussion so far has focused on one particular example, 40 measurements of a continuous variable x, which denoted the range of a projectile fired from a certain gun. We defined the number χ^2 and saw that it is at least a rough measure of the agreement between our observed distribution of measurements and the Gauss distribution that we expected our measurements to follow. As we will now discuss, we can define and use χ^2 in the same way for many different experiments.

Let us consider any experiment in which we measure a number x and for which we have reason to expect a certain distribution of results. We imagine repeating the measurement many times (N) and, having divided the range of possible results x into n bins, $k = 1, \ldots, n$, we count the number O_k of observations that actually fall in each bin k. Assuming the measurements really are governed by the expected distribution, we next calculate the expected number E_k of measurements in the kth bin. Finally, we calculate χ^2 exactly as in (12.5),

$$\chi^2 = \sum_{k=1}^{n} \frac{(O_k - E_k)^2}{E_k}. \tag{12.7}$$

The approximate significance of χ^2 is always the same as in our previous example. That is, if $\chi^2 \lesssim n$, the agreement between our observed and expected distributions is acceptable; if $\chi^2 \gg n$, there is significant disagreement.

The procedure for choosing the bins in terms of which χ^2 is computed depends somewhat on the nature of the experiment being dealt with. Specifically, it depends on whether the measured quantity x is continuous or discrete. We will discuss these two situations in turn.

Measurements of a Continuous Variable

The example discussed in Section 12.1 involved a continuous variable x, and there is little to add to what was said there. The only limiting distribution we have discussed for a continuous variable is the Gauss distribution, but there are, of course, many different distributions that one might expect. For example, in many atomic and nuclear experiments, the expected distribution of the measured variable x (actually an energy) is the Lorentzian distribution

$$f(x) \propto \frac{1}{(x - X)^2 + \gamma^2},$$

where X and γ are certain constants.

Whatever the expected distribution $f(x)$, the total area under the graph of $f(x)$ against x is 1, and the probability of a measurement between $x = a$ and $x = b$ is just the area between a and b,

$$P(a < x < b) = \int_a^b f(x)\,dx.$$

Thus, if the kth bin runs from $x = a_k$ to $x = a_{k+1}$, the expected number of measurements in the kth bin (after N measurements in all) is

$$\begin{aligned} E_k &= N \times P(a_k < x < a_{k+1}) \\ &= N \int_{a_k}^{a_{k+1}} f(x)\,dx. \end{aligned} \tag{12.8}$$

When we discuss the quantitative use of the χ^2 test in Section 12.4, we will see that the expected numbers E_k should not be too small. Although there is no definite lower limit, E_k should probably be about 5 or more,

$$E_k \gtrsim 5. \tag{12.9}$$

We must therefore choose the bins in such a way that E_k as given by (12.8) satisfies this condition. We will also see that the number of bins must not be too small. For instance, in the example of Section 12.1, where the expected distribution was a Gauss distribution whose center X and width σ were not known in advance, the χ^2 test cannot work (as we will see) with less than four bins; that is, in this example we needed to have

$$n \geqslant 4. \tag{12.10}$$

Combining (12.9) and (12.10), we see that we cannot usefully apply the χ^2 test to this kind of experiment if our total number of observations is less than about 20.

Measurement of a Discrete Variable

Suppose now we measure a discrete variable, such as the now familiar number of aces when we throw several dice. In practice, the commonest discrete variable is an integer (like the number of aces), and we shall denote the discrete variable by ν instead of x (which we use for a continuous variable). If we throw five dice, the possible values of ν are $\nu = 0, 1, \ldots, 5$ and it is not actually necessary to group the possible results into bins. We can simply count how many times we got each of the six possible results.

We can express this differently by saying that we have chosen six bins, with each bin containing just one possible result.

Nonetheless, it is often desirable to group several different results into one bin. For instance, if we threw our five dice 200 times, then (according to the probabilities found in Problem 10.7) the expected distribution of results is as shown in the first two columns of Table 12.5. We see that here the expected numbers of throws giving four and five aces are .6 and .03, respectively, both much less than the five or so occurrences required in each bin if we want to use the χ^2 test. This difficulty is easily remedied by grouping the results for $\nu = 3$, 4, and 5 into a single bin. This leaves us with four bins, $k = 1, 2, 3, 4$, which are shown, with the corresponding expected numbers E_k, in the last two columns of Table 12.5.

Table 12.5. Expected occurrence of ν aces ($\nu = 0, 1, \ldots, 5$) when throwing five dice 200 times

Result	Expected occurrences	Bin number k	Expected number E_k
No aces	80.4	1	80.4
One ace	80.4	2	80.4
Two	32.2	3	32.2
Three	6.4		
Four	.6	4	7.0
Five	.03		

Having chosen bins as just described, we could count the observed occurrences O_k in each bin. We could then compute χ^2 and see whether the observed and expected distributions seem to agree. In this experiment we know that the expected distribution is certainly the binomial distribution $b_{5,1/6}(\nu)$ *provided* the dice are true (so that p really is $\frac{1}{6}$). Thus our test of the distribution is, in this case, a test of whether the dice are true or loaded.

In any experiment involving a discrete variable, the bins can be chosen to contain just one result each, provided the expected number of occurrences for each bin is at least the needed five or so. Otherwise, several different results should be grouped together into a single larger bin that does include enough expected occurrences.

Other Forms of χ^2

We have used the notation χ^2 earlier in the book. It was used in Equation (7.6) and again in (8.5); and it could have been used for the sum of squares in (5.42). In all these cases χ^2 is a sum of squares with the general form

$$\chi^2 = \sum_1^n \left(\frac{\text{observed value} - \text{expected value}}{\text{standard deviation}} \right)^2. \tag{12.11}$$

In all cases, χ^2 is an indicator of the agreement between the observed and expected values of some variable. If the agreement is good, χ^2 will be of order n; and if it is bad, χ^2 will be much greater than n.

Unfortunately, we can only use χ^2 to test this agreement if we know the expected values and the standard deviation, and can therefore calculate (12.11). Perhaps the commonest situation in which these are known accurately enough is the kind of test discussed in this chapter; namely, a test of a distribution, with E_k given by the distribution, and the standard deviation given by $\sqrt{E_k}$. Nevertheless, the χ^2 test is of very wide application. Consider, for example, the problem discussed in Chapter 8, the measurement of two variables x and y, where y is expected to be some definite function of x,

$$y = f(x)$$

(like $y = A + Bx$). Suppose we have N measured pairs (x_i, y_i), where the x_i have negligible uncertainty and the y_i have known uncertainties σ_i. Here the expected value of y_i is $f(x_i)$, and we could test how well y fits the function $f(x)$ by calculating

$$\chi^2 = \sum_1^N \left(\frac{y_i - f(x_i)}{\sigma_i} \right)^2.$$

All our previous remarks about the expected value of χ^2 would apply to this number, and the quantitative tests described in the following sections could be used. We will not pursue this important application here, since it is rather rare in the introductory physics laboratory for the uncertainties σ_i to be known reliably enough, but see Problem 12.14.

12.3. *Degrees of Freedom and Reduced χ^2*

We have argued that we can test agreement between an observed and an expected distribution by computing χ^2 and comparing it with the number of bins used in collecting the data. It turns out that a slightly better procedure is to compare χ^2, not with the number of bins n, but instead with the *number of degrees of freedom,* denoted d. We have mentioned the notion of degrees of freedom briefly in Section 8.3. We must now discuss it in more detail.

In general, the number of degrees of freedom d in a statistical calculation is defined as the number of observed data *minus* the number of parameters computed from the data and used in the calculation. For the problems considered in this chapter, the observed data are the numbers of observations O_k in the n bins, $k = 1, \ldots, n$. Thus the number of observed data is just n, the number of bins. Therefore, in the problems considered here,

$$d = n - c,$$

where n is the number of bins and c is the number of parameters that had to be calculated from the data in order to compute the expected numbers E_k. The number c is often called the number of *constraints,* as we will explain shortly.

The number of constraints c varies according to the problem under consideration. Consider first the dice-throwing experiment of Section 12.2. If we throw five dice and are testing the hypothesis that the dice are true, then the expected distribution of numbers of aces is the binomial distribution $b_{5,1/6}(\nu)$, where $\nu = 0, \ldots, 5$ is the number of aces in any one throw. Both parameters in this function—the number of dice, five, and the probability of an ace, 1/6—are known in advance, and do not have to be calculated from the data. When we calculate the expected number of occurrences of any particular ν, we must multiply the binomial probability by the total number of throws N (in our example, $N = 200$). This parameter *does* depend on the data. Specifically, N is just the sum of the numbers O_k,

$$N = \sum_{k=1}^{n} O_k. \tag{12.12}$$

Thus in calculating the expected results of our dice experiment, we have to calculate just one parameter (N) from the data. The number of constraints is therefore

$$c = 1,$$

and the number of degrees of freedom is

$$d = n - 1.$$

In Table 12.5 the results of the dice experiment were grouped into four bins (i.e., $n = 4$); so in that experiment there were 3 degrees of freedom.

The equation (12.12) illustrates well the curious terminology of "constraints" and "degrees of freedom." Once the number N has been determined, one can regard (12.12) as an equation which "constrains" the values of $O_1, \ldots, O_n$. More specifically, we can say that, because of the constraint (12.12), only $n - 1$ of the numbers $O_1, \ldots, O_n$ are independent. For instance, the first $n - 1$ numbers $O_1, \ldots, O_{n-1}$ could take any value (within certain ranges), but then the last number O_n would be completely determined by Equation (12.12). In this sense, only $n - 1$ of the data are *free* to take on independent values; so we say that there are only $n - 1$ independent degrees of freedom.

In the first example in this chapter, the range x of a projectile was measured 40 times ($N = 40$). The results were collected into four bins ($n = 4$) and compared with what we would expect for a Gauss distribution $f_{X,\sigma}(x)$. Here there were *three* constraints and hence only one degree of freedom,

$$d = n - c = 4 - 3 = 1.$$

The first constraint is the same as (12.12): the total number of observations N is the sum of the observations O_k in all the bins. But here there were two more constraints, since (as is usual in this kind of experiment) we did not know in advance the parameters X and σ of the expected Gauss distribution $f_{X,\sigma}(x)$. Thus, before we could calculate the expected numbers E_k, we had to estimate X and σ using the data. There were, therefore, three constraints in all; so in this example

$$d = n - 3. \tag{12.13}$$

Incidentally, this explains why we had to use at least four bins in this experiment. We will see that the number of degrees of freedom must always be one or more; so, from (12.13), it is clear that we had to choose $n \geqslant 4$.

In the examples considered here, there will always be at least one constraint (namely, the constraint $N = \sum O_k$, involving the total number of measurements), and there may be one or two more. Thus the number of degrees of freedom, d, will range from $n - 1$ to $n - 3$ (in our examples). When n is large, the difference between n and d is fairly unimportant, but if n is small (as it often is, unfortunately), there is obviously a significant difference.

Armed with the notion of degrees of freedom, we can now begin to make our χ^2 test more precise. It can be shown (though we will not do so) that the *expected* value of χ^2 is precisely d, the number of degrees of freedom,

$$(\text{expected average value of } \chi^2) = d. \tag{12.14}$$

This important equation does not mean that we really expect to find $\chi^2 = d$ after any one series of measurements. It means instead that if we could repeat our whole series of measurements infinitely many times and compute χ^2 each time, then the average of these values of χ^2 would be d. Nonetheless, even after just *one* set of measurements, a comparison of χ^2 with d is an indicator of the agreement. In particular, if our expected distribution was the *correct* expected distribution, then it is very unlikely χ^2 would be a lot larger then d. Turning this around, if we find $\chi^2 \gg d$, we can assert that it is most unlikely that our expected distribution was correct.

We have *not* proved the result (12.14), but we can see that some aspects of the result are reasonable. For example, since $d = n - c$, we can rewrite (12.14) as

$$(\text{expected average value of } \chi^2) = n - c. \tag{12.15}$$

That is, for any given n, the expected value of χ^2 will be smaller when c is larger (that is, if we calculate more parameters from the data). This is just what we should expect. In the example of Section 12.1, we used the data to calculate the center X and width σ of the expected distribution $f_{X,\sigma}(x)$. Naturally, since X and σ were thus chosen to fit the data, we would expect to find a somewhat better agreement between the observed and expected distributions; that is, these two extra constraints would be expected to reduce the value of χ^2. This is just what (12.15) implies.

The result (12.14) suggests a slightly more convenient way to think about our χ^2 test. We introduce a *reduced chi squared* (or *chi squared per degree of freedom*), which we denote by $\tilde{\chi}^2$ and define as

$$\tilde{\chi}^2 = \chi^2/d. \tag{12.16}$$

Since the expected value of χ^2 is d, we see that the

$$(\text{expected average value of } \tilde{\chi}^2) = 1. \tag{12.17}$$

Thus, whatever the number of degrees of freedom, our test can be simply stated as follows: If we obtain a value of $\tilde{\chi}^2$ of order 1 or less, then we have no reason to doubt our expected distribution; if we obtain a value of $\tilde{\chi}^2$ much larger than one, then it is unlikely that our expected distribution is correct.

12.4. Probabilities for χ^2

Our test for agreement between observed data and their expected distribution is still fairly crude. What we now need is a *quantitative* measure of agreement. In particular, we need some guidance on where to draw the boundary between agreement and disagreement. For example, in the experiment of Section 12.1, we made 40 measurements of a certain range x whose distribution should, we believed, be Gaussian. We collected our data into four bins, and found that $\chi^2 = 1.80$. With three constraints there was only one degree of freedom ($d = 1$); so the reduced chi squared, $\tilde{\chi}^2 = \chi^2/d$, is also 1.80,

$$\tilde{\chi}^2 = 1.80.$$

The question is now: Is a value of $\tilde{\chi}^2 = 1.80$ sufficiently larger than one to rule out our expected Gauss distribution or not?

To answer this question, we begin by supposing that our measurements *were* governed by the expected distribution (a Gaussian, in this example). With this assumption one can calculate the *probability* of obtaining a value of $\tilde{\chi}^2$ as large as, or larger than, our value of 1.80. Here this probability turns out to be

$$P(\tilde{\chi}^2 \geqslant 1.80) \approx 18\%,$$

as we will soon see. That is, if our results were governed by the expected distribution, there would be an 18 percent probability of obtaining a value of $\tilde{\chi}^2$ greater than or equal to the value 1.80 that we actually obtained. In other words, in this experiment a value of $\tilde{\chi}^2$ as large as 1.80 is not at all unreasonable; so we would have no reason (based on this evidence) to reject our expected distribution.

Our general procedure should now be reasonably clear. After completing any series of measurements, we calculate the reduced chi squared, which we now denote by $\tilde{\chi}_o{}^2$ (where the subscript o stands for "obtained," since $\tilde{\chi}_o{}^2$ is the value actually obtained). Next, assuming our measurements do follow the expected distribution, we compute the probability

$$P(\tilde{\chi}^2 \geqslant \tilde{\chi}_o{}^2) \tag{12.18}$$

of finding a value of $\tilde{\chi}^2$ greater than or equal to the value $\tilde{\chi}_o^2$ actually obtained. If this probability is high, then our value $\tilde{\chi}_o^2$ is perfectly acceptable, and there is no reason to reject our expected distribution. If this probability is unreasonably low, then a value of $\tilde{\chi}^2$ as large as our observed $\tilde{\chi}_o^2$ is very unlikely (if our measurements were distributed as expected), and it is correspondingly unlikely that our expected distribution is correct.

As always with statistical tests, we have to decide on the boundary between what is "reasonably probable" and what is not. Two common choices are those already mentioned in connection with correlations. With the boundary at 5 percent, we would say that our observed value $\tilde{\chi}_o^2$ indicates a "significant disagreement" if

$$P(\tilde{\chi}^2 \geqslant \tilde{\chi}_o^2) < 5\%,$$

and we would reject our expected distribution at the "5 percent significance level." If we set the boundary at 1 percent, then we could say that the disagreement is "highly significant" if $P(\tilde{\chi}^2 \geqslant \tilde{\chi}_o^2) < 1$ percent and reject the expected distribution at the "1 percent significance level."

Whatever level we choose as our boundary for rejection, the level chosen should be stated. Perhaps even more important, one should state the probability $P(\tilde{\chi}^2 \geqslant \tilde{\chi}_o^2)$, so that the reader can judge its reasonableness.

The calculation of the probabilities $P(\tilde{\chi}^2 \geqslant \tilde{\chi}_o^2)$ is too complicated to describe in this book. However, the results can be easily tabulated, as in Table 12.6 or in the more complete table in Appendix D. It turns out that the probability of getting any particular values of $\tilde{\chi}^2$ depends on the number of degrees of freedom. Thus we shall write the probability of interest as $P_d(\tilde{\chi}^2 \geqslant \tilde{\chi}_o^2)$ to emphasize its dependence on d.

The usual calculation of the probabilities $P_d(\tilde{\chi}^2 \geqslant \tilde{\chi}_o^2)$ treats the observed numbers O_k as continuous variables that are distributed around their expected values E_k according to a Gauss distribution. In the problems considered here, O_k is a discrete variable, distributed according to the Poisson distribution.[2] Provided all numbers involved are reasonably large, the discrete character of the O_k is unimportant, and the Poisson distribution is well approximated by the Gauss function. Under these conditions the tabulated probabilities $P_d(\tilde{\chi}^2 \geqslant \tilde{\chi}_o^2)$ can be safely used. It is for this reason we have said the bins must be chosen so that the expected count, E_k, in each bin is reasonably large (at least 5 or so). For the same reason, the number of bins should not be too small.

[2] We have argued that finding the number O_k amounts to a counting experiment and hence that O_k should follow a Poisson distribution. If the bin k is too large, then this argument is not strictly correct, since the probability of a measurement in the bin is not much less than 1 (which is one of the conditions for the Poisson distribution, as mentioned in Section 11.1); so we must have a reasonable number of bins.

With these warnings, we now give the calculated probabilities $P_d(\tilde{\chi}^2 \geqslant \tilde{\chi}_o^2)$ for a few representative values of d and $\tilde{\chi}_o^2$ in Table 12.6. The numbers in the left column give six choices of d, the number of degrees of freedom ($d = 1, 2, 3, 5, 10, 15$). Those at the other column heads give possible values of the observed $\tilde{\chi}_o^2$. Each cell in the table therefore shows the percentage probability $P_d(\tilde{\chi}^2 \geqslant \tilde{\chi}_o^2)$ as a function of d and $\tilde{\chi}_o^2$. For example, with ten degrees of freedom ($d = 10$), we see that the probability of obtaining $\tilde{\chi}^2 \geqslant 2$ is 3 percent,

$$P_{10}(\tilde{\chi}^2 \geqslant 2) = 3\%.$$

Thus if we obtained a reduced chi squared of 2 in an experiment with ten degrees of freedom, we could conclude that our observations differed "significantly" from the expected distribution and reject the expected distribution at the 5 percent significance level (though not at the 1 percent level).

Table 12.6. The percentage probability $P_d(\tilde{\chi}^2 \geqslant \tilde{\chi}_o^2)$ of obtaining a value of $\tilde{\chi}^2$ greater than or equal to any particular value $\tilde{\chi}_o^2$, assuming that the measurements concerned really are governed by the expected distribution. Blanks indicate probabilities less than 0.05%.

	$\tilde{\chi}_o^2$												
d	0	.25	.5	.75	1.0	1.25	1.5	1.75	2	3	4	5	6
1	100	62	48	39	32	26	22	19	16	8	5	3	1
2	100	78	61	47	37	29	22	17	14	5	2	.7	.2
3	100	86	68	52	39	29	21	15	11	3	.7	.2	—
5	100	94	78	59	42	28	19	12	8	1	.1	—	—
10	100	99	89	68	44	25	13	6	3	.1	—	—	—
15	100	100	94	73	45	23	10	4	1	—	—	—	—

The probabilities in the second column of Table 12.6 are all 100 percent, since one is always certain to get $\tilde{\chi}^2 \geqslant 0$. As $\tilde{\chi}_o^2$ increases, the probability of getting $\tilde{\chi}^2 \geqslant \tilde{\chi}_o^2$ diminishes, but it does so at a rate that depends on d. Thus, for two degrees of freedom ($d = 2$), $P_d(\tilde{\chi}^2 \geqslant 1)$ is 37 percent; whereas for $d = 15$, $P_d(\tilde{\chi}^2 \geqslant 1)$ is 45 percent. Note that $P_d(\tilde{\chi}^2 \geqslant 1)$ is always appreciable (at least 32 percent, in fact); so a value for $\tilde{\chi}_o^2$ of 1 or less is perfectly reasonable, and never requires one to reject the expected distribution.

The minimum value of $\tilde{\chi}_o^2$ that does require one to question the expected distribution depends on d. For one degree of freedom, we see that $\tilde{\chi}_o^2$ can be as large as 4 before the disagreement becomes "significant" (5 percent level). With two degrees of freedom, the corresponding boundary is $\tilde{\chi}_o^2 = 3$; for $d = 5$, it is closer to 2 ($\tilde{\chi}_o^2 = 2.2$, in fact); and so on.

Armed with the probabilities in Table 12.6 (or in Appendix D), we can now assign a quantitative significance to the value of $\tilde{\chi}_o^2$ obtained in any particular experiment. In Section 12.5 we give some examples.

12.5. *Examples*

We have already analyzed rather completely the example of Section 12.1. In this section we consider three more examples to illustrate the application of the χ^2 test.

Another Example of the Gauss Distribution

The example of Section 12.1 involved a measurement whose results were expected to be normally distributed. The normal, or Gauss, distribution is so common that we consider briefly another example. Suppose an anthropologist is interested in the heights of the natives on a certain island. He suspects that the heights of the adult males should be normally distributed, and measures the heights of a sample of 200 men. Using these measurements, he calculates the mean and standard deviation, and uses these numbers as best estimates for the center X and width parameter σ of the expected normal distribution $f_{X,\sigma}(x)$. He now chooses eight bins, as shown in the left two columns of Table 12.7, and groups his observations, with the results shown in the third column.

Our anthropologist now wishes to check whether these results are consistent with the expected normal distribution $f_{X,\sigma}(x)$. To this end he

Table 12.7. Measurements of the heights of 200 adult males.

Bin number k	Heights in bin	Observed number, O_k	Expected number, E_k
1	less than $X - 1.5\sigma$	14	13.4
2	between $X - 1.5\sigma$ and $X - \sigma$	29	18.3
3	between $X - \sigma$ and $X - 0.5\sigma$	30	30.0
4	between $X - 0.5\sigma$ and X	27	38.3
5	between X and $X + 0.5\sigma$	28	38.3
6	between $X + 0.5\sigma$ and $X + \sigma$	31	30.0
7	between $X + \sigma$ and $X + 1.5\sigma$	28	18.3
8	above $X + 1.5\sigma$	13	13.4

first calculates the probability P_k that any one man have height in any particular bin k (assuming a normal distribution). This is the integral of $f_{X,\sigma}(x)$ between the bin boundaries, and is easily found from the table of integrals in Appendix B. The expected number E_k in each bin is then P_k times the total number of men sampled (200). These numbers are shown in the final column of Table 12.7.

To calculate the expected numbers E_k, he had to use three parameters that were calculated from his data (the total number in the sample and his estimates for X and σ). Thus, although there are eight bins, he had three constraints; so the number of degrees of freedom is $d = 8 - 3 = 5$. A simple calculation using the data of Table 12.7 gives for his reduced chi squared

$$\tilde{\chi}^2 = \frac{1}{d} \sum_{i=1}^{8} \frac{(O_k - E_k)^2}{E_k} = 3.5.$$

Since this value is appreciably larger than 1, we immediately suspect that the islanders' heights do not follow the normal distribution. More specifically, we see from Table 12.6 that, if the islanders' heights were distributed as expected, then the probability $P_5(\tilde{\chi}^2 \geqslant 3.5)$ of obtaining a $\tilde{\chi}^2$ as great as 3.5 or greater is about 0.5 percent. By any standards this is very improbable, and we conclude that it is very unlikely that the islanders' heights are normally distributed. In particular, at the 1 percent (or "highly significant") level, we can reject the hypothesis of a normal distribution of heights.

More Dice

In Section 12.2 we discussed an experiment in which five dice were thrown many times and the numbers of aces in each throw recorded. Suppose we make 200 throws and divide the results into bins as discussed before. Assuming that the dice are true, we can calculate the expected numbers E_k as before. These are shown in the third column of Table 12.8.

Table 12.8. Distribution of numbers of aces in 200 throws of 5 dice.

Bin number k	Results in bin	Expected number, E_k	Observed number, O_k
1	no aces	80.4	60
2	one ace	80.4	88
3	two aces	32.2	39
4	3, 4, or 5 aces	7.0	13

In an actual test, five dice were thrown 200 times and the numbers in the last column of Table 12.8 were observed. To test the agreement between the observed and expected distributions, we simply note that there are three degrees of freedom (four bins minus one constraint) and calculate

$$\tilde{\chi}^2 = \frac{1}{3} \sum_{k=1}^{4} \frac{(O_k - E_k)^2}{E_k} = 4.16.$$

Referring back to Table 12.6 we see that with three degrees of freedom the probability of obtaining a $\tilde{\chi}^2$ this large or larger is about 0.7 percent, *if* the dice are true. We conclude that the dice are almost certainly not true. Comparison of the numbers E_k and O_k in Table 12.8 suggests that at least one die is loaded in favor of the ace.

An Example of the Poisson Distribution

As a final example of the use of the χ^2 test, let us consider an experiment in which the expected distribution is the Poisson distribution. Suppose we arrange a Geiger counter to count the arrival of cosmic-ray particles in a certain region. Suppose further that we count the number of particles arriving in 100 separate one-minute intervals, with the results shown in the first two columns of Table 12.9.

Inspection of the numbers in column two immediately suggests that we group all counts $\nu \geqslant 5$ into a single bin. This choice of six bins

Table 12.9. Numbers of cosmic-ray particles observed in 100 separate one-minute intervals.

Counts ν in one minute	Occurrences	Bin number k	Observations O_k in bin k	Expected number E_k
None	7	1	7	7.5
One	17	2	17	19.4
Two	29	3	29	25.2
Three	20	4	20	21.7
Four	16	5	16	14.1
Five	8	6	11	12.1
Six	1			
Seven	2			
Eight or more	0			
Total	100			

$(k = 1, \ldots, 6)$ is shown in the third column and the corresponding numbers O_k in column four.

The hypothesis we wish to test is that the number ν is governed by a Poisson distribution $p_\mu(\nu)$. Since the expected mean count μ is unknown, we must first calculate the average of our hundred counts. This is easily found to be $\bar{\nu} = 2.59$, which gives us our best estimate for μ. Using this value $\mu = 2.59$, we can calculate the probability $p_\mu(\nu)$ of any particular count ν and hence the expected numbers E_k as shown in the final column.

In calculating the numbers E_k, we used two parameters based on the data, the total number of observations (100) and our estimate of μ ($\mu = 2.59$). (Note that since the Poisson distribution is completely determined by μ, we did not have to estimate the standard deviation σ. Indeed, since $\sigma = \sqrt{\mu}$, our estimate for μ automatically gives us an estimate for σ.) There are, therefore, two constraints, which reduces our six bins to four degrees of freedom, $d = 4$.

A simple calculation using the numbers in the last two columns of Table 12.9 now gives for the reduced chi squared

$$\tilde{\chi}^2 = \frac{1}{d} \sum_{k=1}^{6} \frac{(O_k - E_k)^2}{E_k} = 0.35.$$

Since this value is less than one, we can conclude immediately that the agreement between our observations and the expected Poisson distribution is satisfactory. More specifically, we see from the table in Appendix D that a value of $\tilde{\chi}^2$ as large as 0.35 is very probable; in fact $P_4(\tilde{\chi}^2 \geqslant 0.35) \approx 85$ percent. Thus our experiment gives us absolutely no reason to doubt the expected Poisson distribution.

The value of $\tilde{\chi}^2 = 0.35$ found in this experiment is actually appreciably less than 1, indicating that our observations fitted the Poisson distribution very well. However, this small value does *not* give stronger evidence that our measurements are governed by the expected distribution than would a value $\tilde{\chi}^2 \approx 1$. If the results really are governed by the expected distribution, and if we were to repeat our series of measurements many times, we would expect many different values of $\tilde{\chi}^2$, fluctuating about the average value 1. Thus, if the measurements are governed by the expected distribution, a value of $\tilde{\chi}^2 = 0.35$ is just the result of a large chance fluctuation away from the expected mean value. In no way does it give extra weight to our conclusion that our measurements do seem to follow the expected distribution.

If you have followed these three examples, you should have no difficulty applying the χ^2 test to any problems likely to be found in an elementary physics laboratory. Several further examples can be found in the problems below. You should certainly test your understanding by trying some of them.

Problems

Reminder: An asterisk indicates that the problem is discussed, or its answer given, in the Answers section at the back of the book.

***12.1** (Section 12.1). Each member of a class of 50 students is given a piece of the same metal (or what is said to be the same metal) and told to find its density. From the 50 results the mean $\bar{\rho}$ and standard deviation σ_ρ are calculated, and it is then decided to test whether the results are normally distributed. To this end the measurements are grouped into four bins with boundaries at $\bar{\rho} - \sigma_\rho$, $\bar{\rho}$, and $\bar{\rho} + \sigma_\rho$, with the results shown in Table 12.10.

Table 12.10.

Bin k	Values of ρ in bin	Observations O_k in bin
1	below $\bar{\rho} - \sigma_\rho$	12
2	between $\bar{\rho} - \sigma_\rho$ and $\bar{\rho}$	13
3	between $\bar{\rho}$ and $\bar{\rho} + \sigma_\rho$	11
4	above $\bar{\rho} + \sigma_\rho$	14

Assuming that the measurements were normally distributed, with center $\bar{\rho}$ and width σ_ρ, calculate the number, E_k, of measurements expected in each bin. Hence calculate χ^2. Do the measurements appear to be normally distributed?

12.2 (Section 12.1). In Problem 4.7 were given 30 measurements of a time t, with mean $\bar{t} = 8.15$ sec and standard deviation $\sigma_t = 0.04$ sec. Group the data into four bins with boundaries at $\bar{t} - \sigma_t$, $\bar{t}$, and $\bar{t} + \sigma_t$, and find the observed numbers O_k in each bin $k = 1, 2, 3, 4$. Assuming the measurements were normally distributed with center at $\bar{t}$ and width σ_t, what are the expected numbers E_k in each bin? Calculate χ^2. Is there any reason to doubt that the measurements are normally distributed?

12.3 (Section 12.2). A gambler decides to test a die by throwing it 240 times. Each throw has six possible outcomes ($k = 1, 2, \ldots, 6$, where k is the face showing) and the distribution of his throws is as shown in Table 12.11.

Table 12.11.

Face showing, k	1	2	3	4	5	6
Occurrences, O_k	20	46	35	45	42	52

What are the expected numbers of occurrences E_k, assuming that the die is true? Treating each possible result as a separate bin, compute χ^2. Does it seem likely that the die is loaded?

***12.4** (Section 12.2). Three dice are thrown 400 times and the number of sixes is recorded for each throw, with the results shown in Table 12.12. Assuming the dice are true, calculate the expected numbers E_k for each of the three bins. (The required probabilities are the binomial probabilities discussed in Section 10.2.) Calculate χ^2. Is there any reason to suspect the dice are loaded?

Table 12.12.

Result	Bin k	Occurrences O_k
No sixes	1	217
One six	2	148
Two or three sixes	3	35

***12.5** (Section 12.3).
(a) For each of Problems 12.1 to 12.4, find the number of constraints c and the number of degrees of freedom d.
(b) Suppose that in Problem 12.1 the accepted value ρ_{acc} of the density was known, and we decided to test the hypothesis that the results were governed by a normal distribution centered on ρ_{acc}. For this test how many constraints would there be, and how many degrees of freedom?

***12.6** (Section 12.4). For the data of Problem 12.1, compute the reduced chi squared $\tilde{\chi}^2$. If the measurements were normally distributed, what is the probability of getting a value of $\tilde{\chi}^2$ this large or larger? At the 5 percent significance level, can you reject the hypothesis that the measurements were normally distributed? At the 1 percent level? (See Appendix D for the necessary probabilities.)

12.7 (Section 12.4). In Problem 12.2, can you reject the assumption of normal distribution at either 5 percent or 1 percent levels of significance? (See Appendix D for the necessary probabilities.)

***12.8** (Section 12.5). A pair of dice is thrown 360 times, and the total score is recorded for each throw. The possible totals are 2, 3, . . . , 12 and their numbers of occurrences are as shown in Table 12.13.

Table 12.13.

Total	2	3	4	5	6	7	8	9	10	11	12
Occurrences	6	14	23	35	57	50	44	49	39	27	16

Calculate the probabilities for each total and hence the expected numbers of occurrences (assuming the dice are true). Calculate χ^2, d, and $\tilde{\chi}^2 = \chi^2/d$. Assuming the dice are true, what is the probability of getting a value of $\tilde{\chi}^2$ this large or larger? At the 5 percent level of significance, can you reject the hypothesis that the dice are true? At the 1 percent level? (See Appendix D for the necessary probabilities.)

12.9 (Section 12.5). In Problem 12.3, find the value of $\tilde{\chi}^2$. Can we conclude that the die was loaded, at the 5 percent significance level? At the 1 percent level? (See Appendix D for the necessary probabilities.)

***12.10** (Section 12.5). In Problem 12.4, what is the value of $\tilde{\chi}^2$? If the dice really are true, what is the probability of getting a value of $\tilde{\chi}^2$ this large or larger? Explain whether the evidence suggests that the dice are loaded. (See Appendix D for the necessary probabilities.)

12.11 (Section 12.5). Calculate χ^2 for the data of Problem 11.5, assuming that the observations should follow the Poisson distribution with mean count $\mu = 3$. (Group all values $\nu \geqslant 6$ into a single bin.) How many degrees of freedom are there? (Don't forget that μ was given in advance, and was not calculated from the data.) What is $\tilde{\chi}^2$? Are the data consistent with the expected Poisson distribution? (See Appendix D for the necessary probabilities.)

***12.12** (Section 12.5).

(a) A certain radioactive sample is alleged to produce an average of two decays per minute. To check this, a student measures the numbers of decays in 40 separate one-minute intervals, with the results shown in Table 12.14.

Table 12.14.

Number of decays ν	0	1	2	3	4	5 or more
Times observed	11	12	11	4	2	0

If the decays do follow a Poisson distribution with $\mu = 2$, what numbers would the student expect to observe? (Group all observations with $\nu \geqslant 3$ into a single bin.) Calculate χ^2, d, and $\tilde{\chi}^2 = \chi^2/d$. (Don't forget μ was not calculated from the data.) At the 5 percent significance level, would you reject the hypothesis that the sample follows a Poisson distribution with $\mu = 2$?

(b) The student notices that the actual mean of the results is $\bar{\nu} = 1.35$, and therefore decides to test whether the data fit a Poisson distribution with $\mu = 1.35$. In this case, what are d and $\tilde{\chi}^2$? Are the data consistent with this new hypothesis?

***12.13** (Section 12.5). In Chapter 10 we discussed a test for fit to the binomial distribution. We considered n trials each with two outcomes, success (with probability p) and failure (with probability $1 - p$). We then tested whether the observed number of successes, ν, was compatible with some assumed value of p. As long as the numbers involved are reasonably large, we can treat this same problem with the χ^2 test, with two bins—$k = 1$ for successes and $k = 2$ for failures—and one degree of freedom. In the following, you will use both methods and compare results. With large numbers you will find the agreement is excellent; with small numbers, it is less good, but still good enough that χ^2 is very useful indicator.

(a) A soup manufacturer believes he can introduce a different dumpling into his chicken dumpling soup without affecting the soup's popularity. To test this hypothesis he makes 16 cans, labeled "style X," containing the new dumpling, and 16 cans, labeled "style Y," containing the old dumplings. He then sends one of each type to 16 tasters, and asks them which they prefer. If his hypothesis is correct, we should expect eight tasters to prefer X and eight to prefer Y. In fact, the number who favor X is $\nu = 11$. Calculate χ^2 and the probability of getting a value this large or larger. Does the test indicate a significant difference between the two kinds of dumpling? Now, calculate the corresponding probability exactly, using the binomial distribution, and compare your results. Note that the χ^2 test includes deviations away from the expected numbers in either direction. Therefore, for this comparison you should calculate the "two-tailed" probability, for values of ν that deviate from eight by three or more in either direction; i.e., for $\nu = 11$, $12, \ldots, 16$ *and* $\nu = 5, 4, \ldots, 1$.

(b) Repeat part (a) for the next test, in which the manufacturer makes 400 cans of each style and the number preferring X is 225. (In calculating the binomial probabilities use the Gaussian approximation.)

(c) In part (a) the numbers were small enough that the χ^2 test was fairly crude. (It gave a probability of 14 percent, compared with the correct value 21.0 percent.) With one degree of freedom, we can improve on the χ^2 test a little by using an "adjusted χ^2," defined as

$$\text{adjusted } \chi^2 = \sum_{k=1}^{2} \frac{(|O_k - E_k| - \frac{1}{2})^2}{E_k}.$$

Calculate the adjusted χ^2 for the data of part (a), and show that using this value (instead of the usual χ^2) in the table in Appendix D gives a more accurate approximation.[3]

12.14 (Section 12.5). The χ^2 test can be used to test how well a set of measurements (x_i, y_i) of two variables fits an expected relation $y = f(x)$, provided the uncertainties are known reliably. Suppose y and x are expected to satisfy the linear relation

$$y = f(x) = A + Bx. \tag{12.19}$$

(For instance, y might be the length of a metal rod and x its temperature.) Suppose A and B have been predicted theoretically to have the values $A = 50$ and $B = 6$, and that five measurements of x and y have produced the results shown in Table 12.15.

Table 12.15.

x (negligible uncertainty)	1	2	3	4	5
y (all ± 4)	60	56	71	66	86

The uncertainty quoted on y is the standard deviation; i.e., the five measurements of y all have the same standard deviation $\sigma = 4$. Make a table of the observed and expected values of y_i, and calculate χ^2 as

$$\chi^2 = \sum_{1}^{5} \left(\frac{y_i - f(x_i)}{\sigma} \right)^2.$$

Since no parameters were calculated from the data, there are no constraints and hence there are five degrees of freedom. Calculate $\tilde{\chi}^2$, and use the table in Appendix D to find the probability of obtaining a value of $\tilde{\chi}^2$ this large, assuming y does satisfy (12.19). At the 5 percent level, would you reject the expected relation (12.19)? (If the constants A and B were not known in advance, one could calculate them from the data by the method of least squares. One would then proceed as before, but there would now be only three degrees of freedom.)

[3] We have not *justified* the use of the adjusted χ^2 here, but this example does illustrate its superiority. For more details, see H. L. Alder and E. B. Roessler, *Introduction to Probability and Statistics* (Freeman, 1977) p. 263.

Appendixes

APPENDIX A.

Normal Error Integral, I.

If the measurement of a continuous variable x is subject to many small errors, all of them random, then the expected distribution of results is given by the normal, or Gauss, distribution,

$$f_{X,\sigma}(x) = \frac{1}{\sigma\sqrt{2\pi}} e^{-(x-X)^2/2\sigma^2},$$

where X is the true value of x, and σ the standard deviation.

The integral of the normal distribution function, $\int_a^b f_{X,\sigma}(x)\,dx$, is called the *normal error integral*, and is the probability that a measurement fall between $x = a$ and $x = b$,

$$P(a \leqslant x \leqslant b) = \int_a^b f_{X,\sigma}(x)\,dx.$$

Table A shows this integral for $a = X - t\sigma$ and $b = X + t\sigma$. This gives the probability of a measurement within t standard deviations on either side of X,

$$\begin{aligned} P(\text{within } t\sigma) &= P(X - t\sigma \leqslant x \leqslant X + t\sigma) \\ &= \int_{X-t\sigma}^{X+t\sigma} f_{X,\sigma}(x)\,dx \\ &= \frac{1}{\sqrt{2\pi}} \int_{-t}^{t} e^{-z^2/2}\,dz. \end{aligned}$$

This function is sometimes denoted erf(t), but this notation is also used for a slightly different function.

The probability of a measurement *outside* the same interval can be found by subtraction;

$$P(\text{outside } t\sigma) = 100\% - P(\text{within } t\sigma).$$

For further discussions, see Section 5.4 and Appendix B.

Table A. The percentage probability, $P(\text{within } t\sigma) = \int_{X-t\sigma}^{X+t\sigma} f_{X,\sigma}(x)\,dx$, as a function of t.

$X - t\sigma$ X $X + t\sigma$

t	0.00	0.01	0.02	0.03	0.04	0.05	0.06	0.07	0.08	0.09
0.0	0.00	0.80	1.60	2.39	3.19	3.99	4.78	5.58	6.38	7.17
0.1	7.97	8.76	9.55	10.34	11.13	11.92	12.71	13.50	14.28	15.07
0.2	15.85	16.63	17.41	18.19	18.97	19.74	20.51	21.28	22.05	22.82
0.3	23.58	24.34	25.10	25.86	26.61	27.37	28.12	28.86	29.61	30.35
0.4	31.08	31.82	32.55	33.28	34.01	34.73	35.45	36.16	36.88	37.59
0.5	38.29	38.99	39.69	40.39	41.08	41.77	42.45	43.13	43.81	44.48
0.6	45.15	45.81	46.47	47.13	47.78	48.43	49.07	49.71	50.35	50.98
0.7	51.61	52.23	52.85	53.46	54.07	54.67	55.27	55.87	56.46	57.05
0.8	57.63	58.21	58.78	59.35	59.91	60.47	61.02	61.57	62.11	62.65
0.9	63.19	63.72	64.24	64.76	65.28	65.79	66.29	66.80	67.29	67.78
1.0	68.27	68.75	69.23	69.70	70.17	70.63	71.09	71.54	71.99	72.43
1.1	72.87	73.30	73.73	74.15	74.57	74.99	75.40	75.80	76.20	76.60
1.2	76.99	77.37	77.75	78.13	78.50	78.87	79.23	79.59	79.95	80.29
1.3	80.64	80.98	81.32	81.65	81.98	82.30	82.62	82.93	83.24	83.55
1.4	83.85	84.15	84.44	84.73	85.01	85.29	85.57	85.84	86.11	86.38
1.5	86.64	86.90	87.15	87.40	87.64	87.89	88.12	88.36	88.59	88.82
1.6	89.04	89.26	89.48	89.69	89.90	90.11	90.31	90.51	90.70	90.90
1.7	91.09	91.27	91.46	91.64	91.81	91.99	92.16	92.33	92.49	92.65
1.8	92.81	92.97	93.12	93.28	93.42	93.57	93.71	93.85	93.99	94.12
1.9	94.26	94.39	94.51	94.64	94.76	94.88	95.00	95.12	95.23	95.34
2.0	95.45	95.56	95.66	95.76	95.86	95.96	96.06	96.15	96.25	96.34
2.1	96.43	96.51	96.60	96.68	96.76	96.84	96.92	97.00	97.07	97.15
2.2	97.22	97.29	97.36	97.43	97.49	97.56	97.62	97.68	97.74	97.80
2.3	97.86	97.91	97.97	98.02	98.07	98.12	98.17	98.22	98.27	98.32
2.4	98.36	98.40	98.45	98.49	98.53	98.57	98.61	98.65	98.69	98.72
2.5	98.76	98.79	98.83	98.86	98.89	98.92	98.95	98.98	99.01	99.04
2.6	99.07	99.09	99.12	99.15	99.17	99.20	99.22	99.24	99.26	99.29
2.7	99.31	99.33	99.35	99.37	99.39	99.40	99.42	99.44	99.46	99.47
2.8	99.49	99.50	99.52	99.53	99.55	99.56	99.58	99.59	99.60	99.61
2.9	99.63	99.64	99.65	99.66	99.67	99.68	99.69	99.70	99.71	99.72
3.0	99.73	—	—	—	—	—	—	—	—	—
3.5	99.95	—	—	—	—	—	—	—	—	—
4.0	99.994	—	—	—	—	—	—	—	—	—
4.5	99.9993	—	—	—	—	—	—	—	—	—
5.0	99.99994	—	—	—	—	—	—	—	—	—

APPENDIX B.

Normal Error Integral, II.

In certain calculations, a convenient form of the normal error integral is

$$Q(t) = \int_{X}^{X+t\sigma} f_{X,\sigma}(x)\,dx$$

$$= \frac{1}{\sqrt{2\pi}} \int_0^t e^{-z^2/2}\,dz.$$

(This integral is, of course, just half the integral tabulated in Appendix A.) The probability $P(a \leqslant x \leqslant b)$ of a measurement in any interval $a \leqslant x \leqslant b$ can be found from $Q(t)$ by a single subtraction or addition. For example,

$$P(X + \sigma \leqslant x \leqslant X + 2\sigma) = Q(2) - Q(1).$$

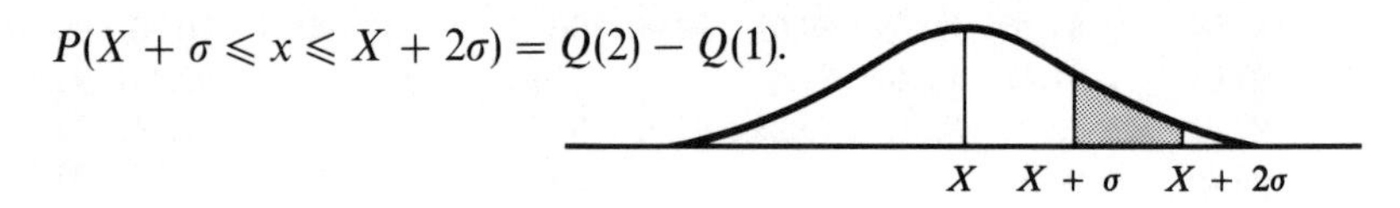

Similarly

$$P(X - 2\sigma \leqslant x \leqslant X + \sigma) = Q(2) + Q(1).$$

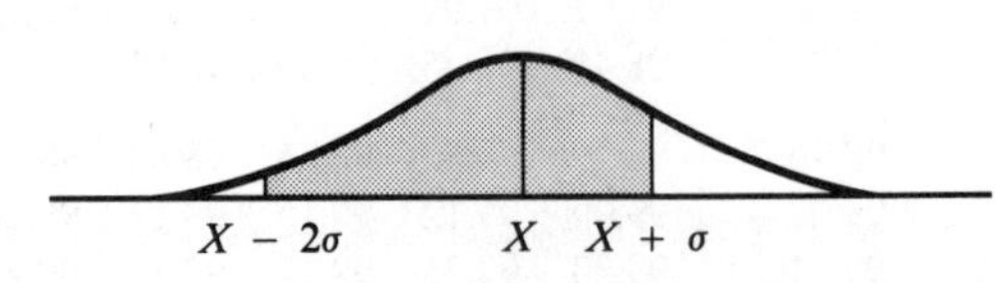

The probability of a measurement greater than any $X + t\sigma$ is just $0.5 - Q(t)$. For example,

$$P(x \geqslant X + \sigma) = 50\% - Q(1).$$

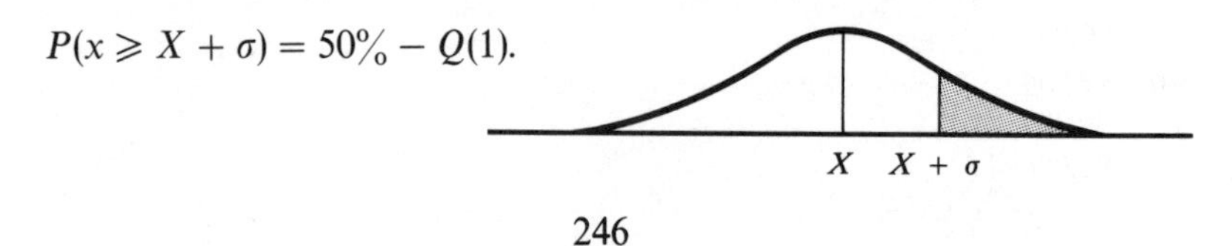

Table B. The percentage probability,
$Q(t) = \int_X^{X+t\sigma} f_{X,\sigma}(x)\,dx$,
as a function of t.

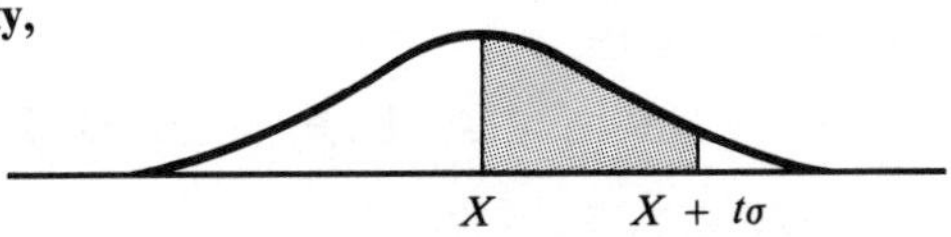

t	0.00	0.01	0.02	0.03	0.04	0.05	0.06	0.07	0.08	0.09
0.0	0.00	0.40	0.80	1.20	1.60	1.99	2.39	2.79	3.19	3.59
0.1	3.98	4.38	4.78	5.17	5.57	5.96	6.36	6.75	7.14	7.53
0.2	7.93	8.32	8.71	9.10	9.48	9.87	10.26	10.64	11.03	11.41
0.3	11.79	12.17	12.55	12.93	13.31	13.68	14.06	14.43	14.80	15.17
0.4	15.54	15.91	16.28	16.64	17.00	17.36	17.72	18.08	18.44	18.79
0.5	19.15	19.50	19.85	20.19	20.54	20.88	21.23	21.57	21.90	22.24
0.6	22.57	22.91	23.24	23.57	23.89	24.22	24.54	24.86	25.17	25.49
0.7	25.80	26.11	26.42	26.73	27.04	27.34	27.64	27.94	28.23	28.52
0.8	28.81	29.10	29.39	29.67	29.95	30.23	30.51	30.78	31.06	31.33
0.9	31.59	31.86	32.12	32.38	32.64	32.89	33.15	33.40	33.65	33.89
1.0	34.13	34.38	34.61	34.85	35.08	35.31	35.54	35.77	35.99	36.21
1.1	36.43	36.65	36.86	37.08	37.29	37.49	37.70	37.90	38.10	38.30
1.2	38.49	38.69	38.88	39.07	39.25	39.44	39.62	39.80	39.97	40.15
1.3	40.32	40.49	40.66	40.82	40.99	41.15	41.31	41.47	41.62	41.77
1.4	41.92	42.07	42.22	42.36	42.51	42.65	42.79	42.92	43.06	43.19
1.5	43.32	43.45	43.57	43.70	43.82	43.94	44.06	44.18	44.29	44.41
1.6	44.52	44.63	44.74	44.84	44.95	45.05	45.15	45.25	45.35	45.45
1.7	45.54	45.64	45.73	45.82	45.91	45.99	46.08	46.16	46.25	46.33
1.8	46.41	46.49	46.56	46.64	46.71	46.78	46.86	46.93	46.99	47.06
1.9	47.13	47.19	47.26	47.32	47.38	47.44	47.50	47.56	47.61	47.67
2.0	47.72	47.78	47.83	47.88	47.93	47.98	48.03	48.08	48.12	48.17
2.1	48.21	48.26	48.30	48.34	48.38	48.42	48.46	48.50	48.54	48.57
2.2	48.61	48.64	48.68	48.71	48.75	48.78	48.81	48.84	48.87	48.90
2.3	48.93	48.96	48.98	49.01	49.04	49.06	49.09	49.11	49.13	49.16
2.4	49.18	49.20	49.22	49.25	49.27	49.29	49.31	49.32	49.34	49.36
2.5	49.38	49.40	49.41	49.43	49.45	49.46	49.48	49.49	49.51	49.52
2.6	49.53	49.55	49.56	49.57	49.59	49.60	49.61	49.62	49.63	49.64
2.7	49.65	49.66	49.67	49.68	49.69	49.70	49.71	49.72	49.73	49.74
2.8	49.74	49.75	49.76	49.77	49.77	49.78	49.79	49.79	49.80	49.81
2.9	49.81	49.82	49.82	49.83	49.84	49.84	49.85	49.85	49.86	49.86
3.0	49.87	—	—	—	—	—	—	—	—	—
3.5	49.98	—	—	—	—	—	—	—	—	—
4.0	49.997	—	—	—	—	—	—	—	—	—
4.5	49.9997	—	—	—	—	—	—	—	—	—
5.0	49.99997	—	—	—	—	—	—	—	—	—

APPENDIX C.

Probabilities for Correlation Coefficients.

The extent to which N points $(x_1, y_1), \ldots, (x_N, y_N)$ fit a straight line is indicated by the linear correlation coefficient

$$r = \frac{\sum(x_i - \bar{x})(y_i - \bar{y})}{[\sum(x_i - \bar{x})^2 \sum(y_i - \bar{y})^2]^{1/2}},$$

which always lies in the interval $-1 \leqslant r \leqslant 1$. Values of r close to ± 1 indicate a good linear correlation; values close to 0 indicate little or no correlation.

A more quantitative measure of the fit can be found by using Table C. For any definite r_o, $P_N(|r| \geqslant |r_o|)$ is the probability that N measurements of two uncorrelated variables would give a coefficient r as large as r_o. Thus if we obtain a coefficient r_o for which $P_N(|r| \geqslant |r_o|)$ is small, then it is correspondingly unlikely that our variables are uncorrelated; that is, a correlation is indicated. In particular, if $P_N(|r| \geqslant |r_o|) \leqslant 5$ percent, the correlation is called *significant*; if it is less than 1 percent, the correlation is called *highly significant.*

For example, the probability that 20 measurements ($N = 20$) of two uncorrelated variables would yield $|r| \geqslant .5$ is given in the table as 2.5 percent. Thus if 20 measurements gave $r = .5$, we would have *significant* evidence of a linear correlation between the two variables. For further discussion, see Sections 9.3 to 9.5.

The values in Table C were calculated from the integral

$$P_N(|r| \geqslant |r_o|) = \frac{2\Gamma[(N-1)/2]}{\sqrt{\pi}\,\Gamma[(N-2)/2]} \int_{|r_o|}^{1} (1 - r^2)^{(N-4)/4}\, dr.$$

See, for example, E. M. Pugh and G. H. Winslow, *The Analysis of Physical Measurements* (Addison-Wesley, 1966), Section 12-8.

Table C. The percentage probability $P_N(|r| \geqslant r_o)$ that N measurements of two uncorrelated variables give a correlation coefficient with $|r| \geqslant r_o$, as a function of N and r_o. (Blanks indicate probabilities less than 0.05 percent.)

	r_o										
N	0	.1	.2	.3	.4	.5	.6	.7	.8	.9	1
3	100	94	87	81	74	67	59	51	41	29	0
4	100	90	80	70	60	50	40	30	20	10	0
5	100	87	75	62	50	39	28	19	10	3.7	0
6	100	85	70	56	43	31	21	12	5.6	1.4	0
7	100	83	67	51	37	25	15	8.0	3.1	0.6	0
8	100	81	63	47	33	21	12	5.3	1.7	0.2	0
9	100	80	61	43	29	17	8.8	3.6	1.0	0.1	0
10	100	78	58	40	25	14	6.7	2.4	0.5	—	0
11	100	77	56	37	22	12	5.1	1.6	0.3	—	0
12	100	76	53	34	20	9.8	3.9	1.1	0.2	—	0
13	100	75	51	32	18	8.2	3.0	0.8	0.1	—	0
14	100	73	49	30	16	6.9	2.3	0.5	0.1	—	0
15	100	72	47	28	14	5.8	1.8	0.4	—	—	0
16	100	71	46	26	12	4.9	1.4	0.3	—	—	0
17	100	70	44	24	11	4.1	1.1	0.2	—	—	0
18	100	69	43	23	10	3.5	0.8	0.1	—	—	0
19	100	68	41	21	9.0	2.9	0.7	0.1	—	—	0
20	100	67	40	20	8.1	2.5	0.5	0.1	—	—	0
25	100	63	34	15	4.8	1.1	0.2	—	—	—	0
30	100	60	29	11	2.9	0.5	—	—	—	—	0
35	100	57	25	8.0	1.7	0.2	—	—	—	—	0
40	100	54	22	6.0	1.1	0.1	—	—	—	—	0
45	100	51	19	4.5	0.6	—	—	—	—	—	0
	0	.05	.1	.15	.2	.25	.3	.35	.4	.45	
50	100	73	49	30	16	8.0	3.4	1.3	0.4	0.1	
60	100	70	45	25	13	5.4	2.0	0.6	0.2	—	
70	100	68	41	22	9.7	3.7	1.2	0.3	0.1	—	
80	100	66	38	18	7.5	2.5	0.7	0.1	—	—	
90	100	64	35	16	5.9	1.7	0.4	0.1	—	—	
100	100	62	32	14	4.6	1.2	0.2	—	—	—	

APPENDIX D.

Probabilities for χ^2

If a series of measurements is grouped into bins $k = 1, \ldots, n$, we denote by O_k the number of measurements observed in the bin k. The number *expected* (on the basis of some assumed or expected distribution) in the bin k is denoted by E_k. The extent to which the observations fit the assumed distribution is indicated by the reduced chi squared, $\tilde{\chi}^2$, defined as

$$\tilde{\chi}^2 = \frac{1}{d} \sum_{k=1}^{n} \frac{(O_k - E_k)^2}{E_k},$$

where d is the number of degrees of freedom, $d = n - c$, and c is the number of constraints (see Section 12.3). The expected average value of $\tilde{\chi}^2$ is 1. If $\tilde{\chi}^2 \gg 1$, the observed results do not fit the assumed distribution; if $\tilde{\chi}^2 \lesssim 1$, the agreement is satisfactory.

This test is made quantitative with the probabilities shown in Table D. Let $\tilde{\chi}_o{}^2$ denote the value of $\tilde{\chi}^2$ actually obtained in an experiment with d degrees of freedom. The number $P_d(\tilde{\chi}^2 \geqslant \tilde{\chi}_o{}^2)$ is the probability of obtaining a value of $\tilde{\chi}^2$ as large as the observed $\tilde{\chi}_o{}^2$, if the measurements really did follow the assumed distribution. Thus if $P_d(\tilde{\chi}^2 \geqslant \tilde{\chi}_o{}^2)$ is large, the observed and expected distributions are consistent; if it is small, they probably disagree. In particular, if $P_d(\tilde{\chi}^2 \geqslant \tilde{\chi}_o{}^2)$ is less than 5 percent, we say the disagreement is *significant*, and reject the assumed distribution at the 5 percent level. If it is less than 1 percent, the disagreement is called *highly significant*, and we reject the assumed distribution at the 1 percent level.

For example, suppose that we obtain a reduced chi squared of 2.6 (i.e., $\tilde{\chi}_o{}^2 = 2.6$) in an experiment with six degrees of freedom ($d = 6$). According to Table D, the probability of getting $\tilde{\chi}^2 \geqslant 2.6$ is 1.6 percent, if the measurements were governed by the assumed distribution. Thus at the 5 percent level (but not quite at the 1 percent level), we would reject the assumed distribution. For further discussion see Chapter 12.

Table D. The percentage probability $P_d(\tilde{\chi}^2 \geqslant \tilde{\chi}_o^2)$ of obtaining a value of $\tilde{\chi}^2 \geqslant \tilde{\chi}_o^2$ in an experiment with d degrees of freedom, as a function of d and $\tilde{\chi}_o^2$. (Blanks indicate probabilities less than 0.05 percent.)

	$\tilde{\chi}_o^2$														
d	0	0.5	1.0	1.5	2.0	2.5	3.0	3.5	4.0	4.5	5.0	5.5	6.0	8.0	10.0
1	100	48	32	22	16	11	8.3	6.1	4.6	3.4	2.5	1.9	1.4	0.5	0.2
2	100	61	37	22	14	8.2	5.0	3.0	1.8	1.1	0.7	0.4	0.2	—	—
3	100	68	39	21	11	5.8	2.9	1.5	0.7	0.4	0.2	0.1	—	—	—
4	100	74	41	20	9.2	4.0	1.7	0.7	0.3	0.1	0.1	—	—	—	—
5	100	78	42	19	7.5	2.9	1.0	0.4	0.1	—	—	—	—	—	—

d	0	0.2	0.4	0.6	0.8	1.0	1.2	1.4	1.6	1.8	2.0	2.2	2.4	2.6	2.8	3.0
1	100	65	53	44	37	32	27	24	21	18	16	14	12	11	9.4	8.3
2	100	82	67	55	45	37	30	25	20	17	14	11	9.1	7.4	6.1	5.0
3	100	90	75	61	49	39	31	24	19	14	11	8.6	6.6	5.0	3.8	2.9
4	100	94	81	66	52	41	31	23	17	13	9.2	6.6	4.8	3.4	2.4	1.7
5	100	96	85	70	55	42	31	22	16	11	7.5	5.1	3.5	2.3	1.6	1.0
6	100	98	88	73	57	42	30	21	14	9.5	6.2	4.0	2.5	1.6	1.0	0.6
7	100	99	90	76	59	43	30	20	13	8.2	5.1	3.1	1.9	1.1	0.7	0.4
8	100	99	92	78	60	43	29	19	12	7.2	4.2	2.4	1.4	0.8	0.4	0.2
9	100	99	94	80	62	44	29	18	11	6.3	3.5	1.9	1.0	0.5	0.3	0.1
10	100	100	95	82	63	44	29	17	10	5.5	2.9	1.5	0.8	0.4	0.2	0.1
11	100	100	96	83	64	44	28	16	9.1	4.8	2.4	1.2	0.6	0.3	0.1	0.1
12	100	100	96	84	65	45	28	16	8.4	4.2	2.0	0.9	0.4	0.2	0.1	—
13	100	100	97	86	66	45	27	15	7.7	3.7	1.7	0.7	0.3	0.1	0.1	—
14	100	100	98	87	67	45	27	14	7.1	3.3	1.4	0.6	0.2	0.1	—	—
15	100	100	98	88	68	45	26	14	6.5	2.9	1.2	0.5	0.2	0.1	—	—
16	100	100	98	89	69	45	26	13	6.0	2.5	1.0	0.4	0.1	—	—	—
17	100	100	99	90	70	45	25	12	5.5	2.2	0.8	0.3	0.1	—	—	—
18	100	100	99	90	70	46	25	12	5.1	2.0	0.7	0.2	0.1	—	—	—
19	100	100	99	91	71	46	25	11	4.7	1.7	0.6	0.2	0.1	—	—	—
20	100	100	99	92	72	46	24	11	4.3	1.5	0.5	0.1	—	—	—	—
22	100	100	99	93	73	46	23	10	3.7	1.2	0.4	0.1	—	—	—	—
24	100	100	100	94	74	46	23	9.2	3.2	0.9	0.3	0.1	—	—	—	—
26	100	100	100	95	75	46	22	8.5	2.7	0.7	0.2	—	—	—	—	—
28	100	100	100	95	76	46	21	7.8	2.3	0.6	0.1	—	—	—	—	—
30	100	100	100	96	77	47	21	7.2	2.0	0.5	0.1	—	—	—	—	—

The values in Table D were calculated from the integral

$$P_d(\tilde{\chi}^2 \geqslant \tilde{\chi}_o^{\ 2}) = \frac{2}{2^{d/2}\Gamma(d/2)} \int_{\chi_o}^{\infty} x^{d-1} e^{-x^2/2}\, dx.$$

See, for example, E. M. Pugh and G. H. Winslow, *The Analysis of Physical Measurements* (Addison-Wesley, 1966) Section 12-5.

Bibliography

The following are books that I have found useful. They are arranged approximately according to their mathematical level and the completeness of their coverage.

A beautifully clear introduction to statistical methods that contrives to use no calculus is Oliver L. Lacy, *Statistical Methods in Experimentation* (MacMillan, 1953).

A more advanced book on statistics that is also very clear and uses no calculus is Henry L. Alder and Edward B. Roessler, *Introduction to Probability and Statistics* (Freeman, 1977).

Three books at approximately the level of this book that cover many of the same topics are:

D. C. Baird, *Experimentation; An Introduction to Measurement Theory and Experiment Design* (Prentice Hall, 1962);

N. C. Barford, *Experimental Measurements; Precision, Error, and Truth* (Addison-Wesley, 1967);

Hugh D. Young, *Statistical Treatment of Experimental Data* (McGraw-Hill, 1962).

Numerous further topics and derivations can be found in the following more advanced books:

Philip R. Bevington, *Data Reduction and Error Analysis for the Physical Sciences* (McGraw-Hill, 1969);

Stuart L. Meyer, *Data Analysis for Scientists and Engineers* (John Wiley, 1975);

Emerson M. Pugh and George H. Winslow, *The Analysis of Physical Measurements* (Addison-Wesley, 1966).

Answers to Selected Problems

Note on significant figures: Small differences in the least significant digit can arise from use of different rounding procedures, and these are usually unimportant. For the problems of Chapters 2 and 3, the uncertainties given below were found by the crudest method possible, by rounding to one significant figure at each stage of the calculation. In the few cases where a more exact procedure gives a different result, the exact answer is shown in parentheses, suitably rounded at the end of the calculation. In Chapters 4-12, all answers were found with a calculator (which carries ten digits) and rounded afterward.

Chapter 2

2.2 (a) $5.03 \pm .04$ meters. (b) This is a strong case for keeping an extra digit and giving 19.5 ± 1 sec. (c) $(-3.2 \pm .3) \times 10^{-19}$ coulombs. (d) $(.56 \pm .07) \times 10^{-6}$ meters. (e) $(3.27 \pm .04) \times 10^3$ gm·cm/sec.

2.3 (a) Probably the only reasonable conclusion at this stage is $1.9 \pm .1$ gm/cm^3. (b) The discrepancy is .05 gm/cm^3, which is not significant.

2.5 The column headed $(L - L')$ should read: $.3 \pm .9$, $-.6 \pm 1.5$, -2.2 ± 2 (which could be rounded to -2 ± 2), 1 ± 4, 1 ± 4, -4 ± 4. The difference $(L - L')$ should theoretically be zero. In all cases except one the measured value is smaller than its uncertainty; and in the one exceptional case (-2.2 ± 2) it is only slightly larger. Thus the observed values are consistent with the expected value zero.

2.8 (a) Since a straight line can be found (as in Figure A2.8) that passes through 0 and through all the error bars, the data *are* consistent with the prediction $v^2 \propto h$.
(b) Slope of best fit ≈ 18.4; slope of steepest reasonable fit ≈ 20.4; slope of least steep reasonable fit ≈ 16.4. Thus slope $= 18.4 \pm 2$ m/sec^2 (or perhaps 18 ± 2), which is consistent with the expected value 19.6 m/sec^2. Question: When one draws such lines, should one insist that they pass through 0 or not? The answer depends on

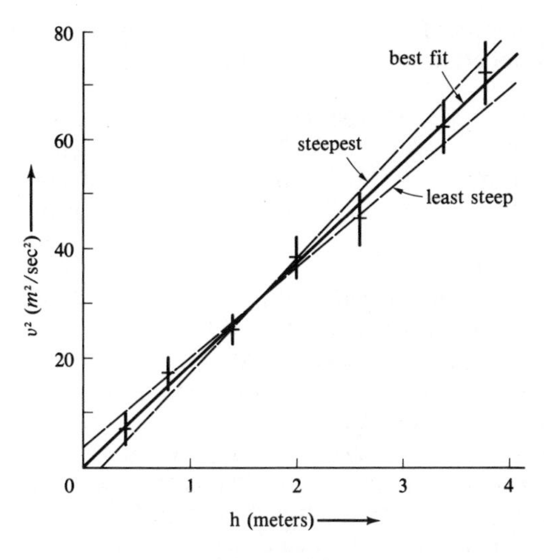

Figure A2.8.

the details of the measurements. Here we have allowed the lines to miss 0, and hence have a generous estimate of the uncertainty.

2.9 (a) In Figure A2.9(a), which includes the origin, it is impossible to say whether T varies with A. From Figure A2.9(b), with its much enlarged vertical scale, it is clear that T does vary with A. Obviously one must consider carefully what is the best choice of axes for the purpose at hand.

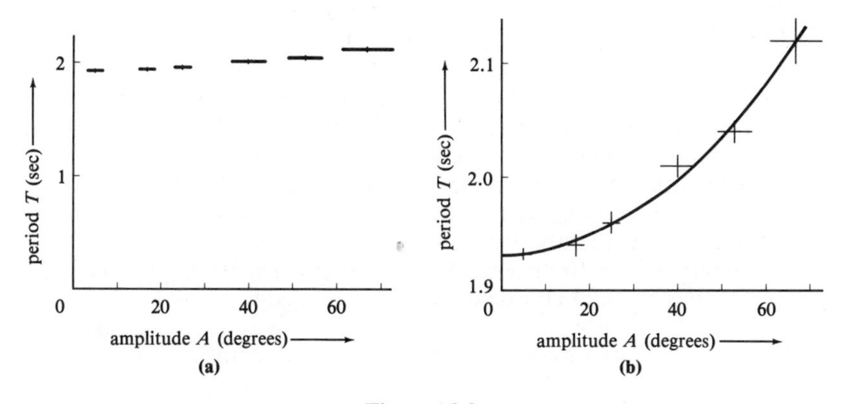

Figure A2.9.

(b) If either picture were redrawn with error bars of 0.3 sec (up and down), it would be clear that here there is no evidence for variation of T with A.

2.12 (a) The answers for $v_f - v_i$ are 4.0 ± .3 and .6 ± .4 cm/sec. (b) The percent uncertainties are 8 percent and 70 percent.

2.14

	Answer	*Percentage uncertainty*	*Absolute uncertainty*
(a)	292 cm^2	3%	9 cm^2 (or 7 "exactly")
(b)	270 cm·sec	10%	30 cm·sec
(c)	12 ft·lb	10%	1 ft·lb

2.15 (a) $q_{best} = 10 \times 20 = 200$;
(highest probable value of q) = $11 \times 21 = 231$;
(lowest probable value of q) = $9 \times 19 = 171$.
The rule in (2.27) gives $q = 200 \pm 30$, which agrees well.
(b) (highest probable value) = $18 \times 35 = 630$;
(lowest probable value) = $2 \times 5 = 10$.
The rule in (2.27) gives $q = 200 \pm 300$ (i.e., $q_{max} = 500$; $q_{min} = -100$). The reason this is so badly incorrect is that the rule in (2.27) applies only when the fractional uncertainties are small compared to one. This condition (which *is* usually met in practice) is not met here.

Chapter 3

3.1 (a) $32 \pm \sqrt{32} \approx 32 \pm 6$. (b) $786 \pm \sqrt{786} \approx 790 \pm 30$. (c) 16 ± 3 for A, 13.1 ± 0.5 for B. Note that A's and B's answers are consistent, but that B has been rewarded with a smaller uncertainty.
3.3 (a) 3 ± 7. (b) 40 ± 20. (c) 0.5 ± 0.1. (d) 63 ± 6.
3.4 (a) 0.48 ± .02 sec (or 4%). (b) 0.470 ± .005 sec (or 1%). (c) No. In the first place the pendulum will eventually stop, unless it is driven. Even if it is driven, other effects will eventually become important and thwart our quest for greater and greater accuracy. For example, if we time for several hours, the reliability of the stopwatch may become a limiting factor, and the period τ may *vary* because of changing temperature, humidity, etc.
3.6 Depth = 40 ± 10 meters. (A more exact calculation gives 44 ± 15, which one might choose to leave unrounded.)

3.8

	"errors add"	"errors add in quadrature"
$a + b$	80 ± 8	80 ± 6
$a + c$	90 ± 6	90 ± 5
$a + d$	58 ± 5	58 ± 5

3.10 (a) .70 ± .05 MeV. (b) .40 ± .02 MeV.
3.11 (a) $\sin\theta = .82 \pm .02$. (Don't forget that $\delta\theta$ has to be expressed in

radians when using $\delta(\sin\theta) = |\cos\theta|\,\delta\theta$.) (b) $f_{\text{best}} = e^{a_{\text{best}}}$, $\delta f = f_{\text{best}}\delta a$, $e^a = 20 \pm 2$. (c) $f_{\text{best}} = \ln a_{\text{best}}$, $\delta f = \delta a/a_{\text{best}}$, $\ln a = 1.10 \pm .03$.

3.14 $n = 1.66 \pm 20\%$, $1.52 \pm 9\%$, $1.54 \pm 6\%$, $1.58 \pm 3\%$, $1.53 \pm 2\%$. As the angle increases, $\delta n/n$ decreases, mainly because the absolute uncertainties are constant; so the fractional uncertainties are smaller when the angles are large.

3.16 (a) 1 and 1. (b) y and x. (c) $2xy^3$ and $3x^2y^2$.

3.17 (c) $\text{LHS} = (x+u)^2(y+v)^3 = (x^2+2xu+u^2)(y^3+3y^2v+3yv^2+v^3)$
$= x^2y^3 + 2xy^3u + 3x^2y^2v +$ (terms involving u^2, uv, v^2, and higher powers).

$\text{RHS} = x^2y^3 + 2xy^3u + 3x^2y^2v.$

Therefore LHS $\approx$ RHS, when u and v are small.

3.19 (a) The correct answer is $\delta q = 0.005$, but step-by-step calculation gives $\delta q = 0.1$. (b) $\delta q = 0.1$ either way. In (a) the numbers are such that a small error in x changes $x + y$ and $x + z$ by almost the same proportion and so cancels in $(x + y)/(x + z)$; the step-by-step calculation ignores this cancelation. In (b) an error in x makes $x + y$ larger but $x + z$ smaller, or vice versa, and so does not cancel out of q.

Chapter 4

4.1 $\bar{x} = 7.2$; $\sigma_x = 1.5$ using definition (4.9), or 1.3 using (4.6).

4.3 $\bar{d} = (1/N)\sum d_i = (1/N)\sum(x_i - \bar{x}) = (1/N)\sum x_i - (1/N)N\bar{x} = \bar{x} - \bar{x} = 0$. If any of these steps is not clear, write out the sums explicitly as $\sum d_i = d_1 + d_2 + \cdots + d_N$, etc.

4.4

$$\begin{aligned}\sum(x_i - \bar{x})^2 &= \sum(x_i^2 - 2\bar{x}x_i + \bar{x}^2)\\ &= \sum x_i^2 - 2\bar{x}\sum x_i + N\bar{x}^2\\ &= \sum x_i^2 - 2\bar{x}N\bar{x} + N\bar{x}^2 = \sum x_i^2 - N\bar{x}^2.\end{aligned}$$

(Again, write out the sums explicitly if you have any doubts.)

4.7 (a) $\bar{t} = 8.149$ sec, $\sigma_t = .039$ sec. (b) Outside $\bar{t} \pm \sigma_t$ we expect 30% or 9 measurements, and we got 8. Outside $\bar{t} \pm 2\sigma_t$ we expect 5% or 1.5, and we got 2.

4.9 (Final answer for t) $= \bar{t} \pm \sigma_{\bar{t}} = 8.149 \pm .007$ sec.

4.11 $\bar{A} = 1221.2\ \text{mm}^2$, $\sigma_{\bar{A}} = .3\ \text{mm}^2$. These compare well with the answer $1221.2 \pm .4\ \text{mm}^2$ found in the text.

4.13 (a) 336 ± 15 m/sec. The 1% systematic uncertainty in f is negligible beside the 4.5% uncertainty in λ. (b) 336 ± 11 m/sec. Here the systematic uncertainty dominates.

Chapter 5

5.1 See Figure A5.1. The broken curve on Figure A5.1(c) is the Gauss function for Problem 5.4.

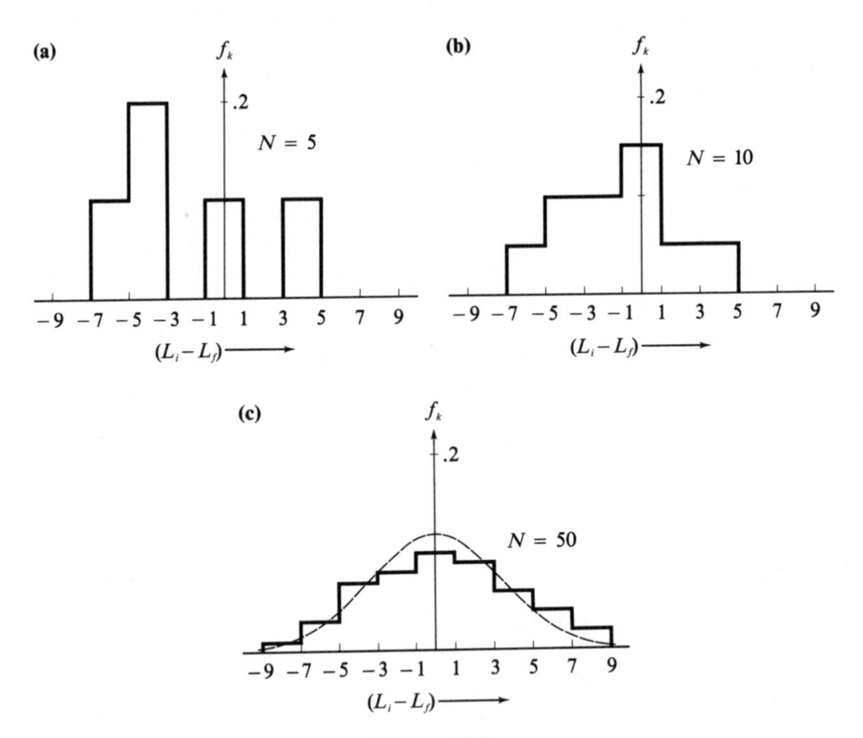

Figure A5.1.

5.2 (a) $C = 1/(2a)$. (b) All values between $-a$ and a are equally likely; no measurements fall outside the range $-a$ to a. (c) $\bar{x} = 0$, $\sigma_x = a/\sqrt{3}$.

5.4 See answers to Problem 5.1.

5.6 The integral $\int z^2 e^{-z^2/2}\,dz$ can be rewritten in the standard form $\int u\,dv$ with $u = z$ and $v = e^{-z^2/2}$. When you integrate by parts, the end-point term $[uv]_{-\infty}^{\infty}$ is zero in this case.

5.8 (a) 68%. (b) 38%. (c) 95%. (d) 48%. (e) 14%. (f) $22.3 \leqslant y \leqslant 23.7$.

5.10 Be careful to differentiate P correctly; you should get

$$\partial P/\partial\sigma = \sigma^{-(N+3)}[\textstyle\sum(x_i - X)^2 - N\sigma^2]\exp[-\textstyle\sum(x_i - X)^2/2\sigma^2];$$

P is maximum when $\partial P/\partial\sigma = 0$, which leads to the desired answer.

5.12 (a) $\sigma_t = 7.04$. (b) $\bar{t}_1 = 74.25$, $\bar{t}_2 = 67.75$, etc. If $\bar{t}$ denotes the average of any group of four measurements, then we would expect $\sigma_{\bar{t}} = \sigma_t/\sqrt{4} = 3.52$; in fact, the SD of the ten means is 3.56. (c) See Figure A5.12.

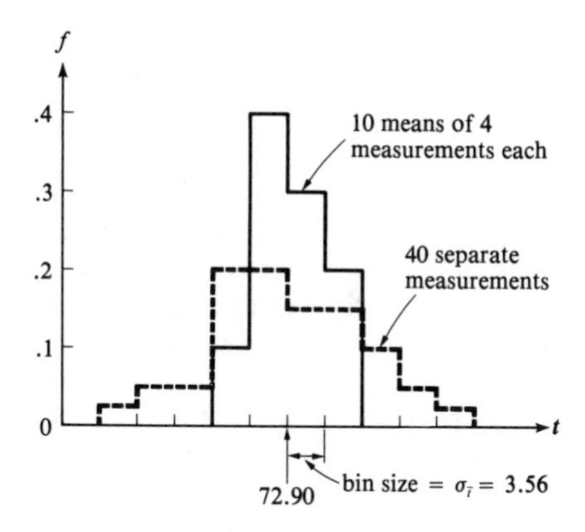

Figure A5.12

5.13 The student's answer (9.5) differs from the supposed center of the distribution (9.8) by 0.3 or three standard deviations. The probability of a result three or more standard deviations from the center is $P(\text{outside } 3\sigma) = 0.3$ percent. This is so improbable that we must suspect his measurements are *not* normally distributed about 9.8 with $\sigma = 0.1$; i.e., either he made some downright mistake, or his experiment was subject to some systematic error.

5.15 $E_f - E_i = 15$ MeV, with a standard deviation of 9.5 MeV. If the measurement was normally distributed, with center $E_f - E_i = 0$ and $\sigma = 9.5$ MeV, then the measurement differs from the true value by 15/9.5, or 1.6, standard deviations. Since $P(\text{outside } 1.6\sigma) =$ 11 percent, the answer is perfectly reasonable and there is no reason to doubt energy conservation.

Chapter 6

6.2 (a) $\bar{V} = .862$ volts, $\sigma_V = .039$ volts. (b) He will reject it. The result .95 differs from $\bar{V}$ by .088 or 2.3σ. Since $P(\text{outside } 2.3\sigma) =$ 2.1 percent, in ten measurements we would expect only 0.21 measurements that differ by this much or more from $\bar{V}$. By Chauvenet's criterion the result must be rejected.

6.3 She does not reject the result 12. Here $\bar{T} = 7.00$ and $\sigma_T = 2.72$; so 12 differs from $\bar{T}$ by 5 or 1.84σ. Since $P(\text{outside } 1.84\sigma) = 6.6$ percent, in 14 measurements we would expect 0.92 measurements that differ this much or more from $\bar{T}$.

Chapter 7

7.1 (a) The two measurements are consistent, and the best estimate based on both is 334.4 ± 0.9 m/sec. (b) These are also consistent (more so, in fact). The best estimate here is 334.08 ± 0.98, which one would almost certainly round to 334 ± 1 m/sec. Evidently the second result is so much more uncertain that it is not worth including.

7.2 (a) 76 ± 4 ohms. (b) About 26 measurements.

7.5 According to (3.47),

$$(\sigma_{x_{\text{best}}})^2 = \sum_i \left(\frac{\partial x_{\text{best}}}{\partial x_i} \sigma_{x_i} \right)^2.$$

The required derivative is $\partial x_{\text{best}}/\partial x_i = w_i/(\sum w_i)$. If you have trouble seeing this, write out the sum $\sum w_i x_i$ as $w_1 x_1 + \cdots + w_N x_N$, and then differentiate with respect to x_1, x_2, etc. Therefore

$$(\sigma_{x_{\text{best}}})^2 = \frac{1}{(\sum w_i)^2} \sum (w_i \sigma_{x_i})^2$$

or, since $\sigma_{x_i} = 1/\sqrt{w_i}$, $(\sigma_{x_{\text{best}}})^2 = 1/\sum w_i$.

Chapter 8

8.1 $A = 9.00$, $B = 2.60$. This gives the solid line in Figure A8.1. (The dashed line is for Problem 8.9.)

8.3 The argument parallels closely that leading from (8.2) to (8.12) in the text, the only important change being that $A = 0$ throughout. Thus $P(y_1, \ldots, y_N) \propto \exp(-\chi^2/2)$ as in (8.4) and χ^2 is given by (8.5), except that $A = 0$. Differentiation with respect to B gives (8.7) (again with $A = 0$), and the solution is $B = (\sum x_i y_i)/(\sum x_i^2)$.

8.4 As in Problem 8.3, the argument closely parallels that leading from Equation (8.2) to (8.12). As in Equation (8.4), $P(y_1, \ldots, y_N) \propto \exp(-\chi^2/2)$, but because the measurements have different uncer-

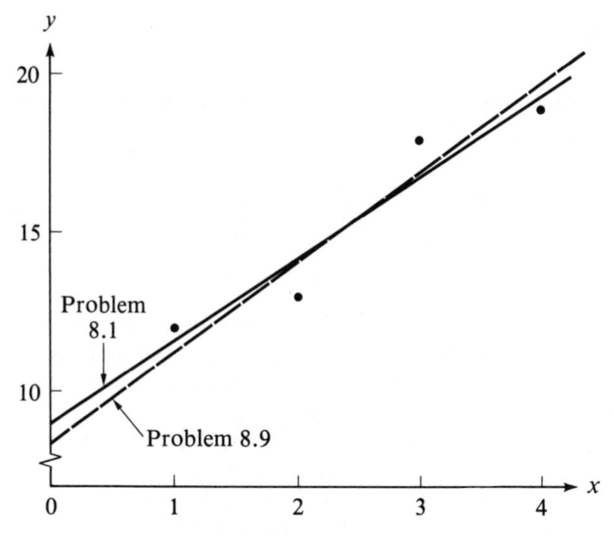

Figure A8.1.

tainties $\chi^2 = \sum w_i(y_i - A - Bx_i)^2$. (Remember, $w_i = 1/\sigma_i{}^2$.) The argument then continues as before.

8.6 $v = 122 \pm 3$ ft/sec.

8.8 (a) You must find the value of σ for which $P(y_1, \ldots, y_N)$ in (8.4) is largest. The derivative $\partial P/\partial\sigma$ is

$$\sigma^{-(N+3)}[\sum(y_i - A - Bx_i)^2 - N\sigma^2] \exp(-\chi^2/2);$$

setting this equal to zero gives the desired value of σ.
(b) The constants A and B are definite functions of $x_1, \ldots, x_N$ and $y_1, \ldots, y_N$. Since the x_i have no uncertainty, the error-propagation formula (3.47) gives, for example,

$$\sigma_A{}^2 = \sum_i \left(\frac{\partial A}{\partial y_i}\sigma_{y_i}\right)^2.$$

Substituting $\partial A/\partial y_i = [(\sum x_i^2) - x_i(\sum x_i)]/\Delta$ gives (8.15) after a little algebra. A similar calculation gives σ_B.

8.9 $A' = -2.9 \pm 1.2$, $B' = 0.35 \pm 0.08$. Using the constants A and B from Problem 8.1, we would find $A' = -3.5$ and $B' = 0.38$; these are within the uncertainties of the new answers. Thus although the two methods do give different lines (see Figure A8.1), the difference is not really significant.

8.11 Best estimate for g is 9.4 m/sec^2.

8.13 $A = 5.5$ cm, $B = 11.1$ cm.

8.14 $\tau = 2.0$ hours.

Chapter 9

9.1 This calculation is simplest if you note that the function $A(t)$ is just $A(t) = \sigma_x{}^2 + 2t\sigma_{xy} + t^2\sigma_y{}^2$.

9.3 (a)
$$\begin{aligned}\textstyle\sum(x_i - \bar{x})(y_i - \bar{y}) &= \textstyle\sum(x_iy_i - \bar{x}y_i - \bar{y}x_i + \bar{x}\bar{y}) \\ &= (\textstyle\sum x_iy_i) - \bar{x}(\sum y_i) - \bar{y}(\sum x_i) + N\bar{x}\bar{y} \\ &= (\textstyle\sum x_iy_i) - N\bar{x}\bar{y}.\end{aligned}$$

9.5 (a) $P_5(|r| \geqslant 0.7) = 19$ percent. Thus after five measurements, a value of $r = 0.7$ is quite likely even if K and f are not linearly correlated. In particular, it does not give "significant" support to a linear relation.
(b) $P_{20}(|r| \geqslant 0.5) = 2$ percent. Since this is less than 5 percent, it gives "significant" evidence for a linear relation.

9.6 (a) $r = -0.97$. Since $P_5(|r| \geqslant 0.97) \approx 1.2$ percent, there is a "significant" correlation. (b) $r = -0.57$. Since $P_5(|r| \geqslant 0.57) \approx 31$ percent, this is not significant.

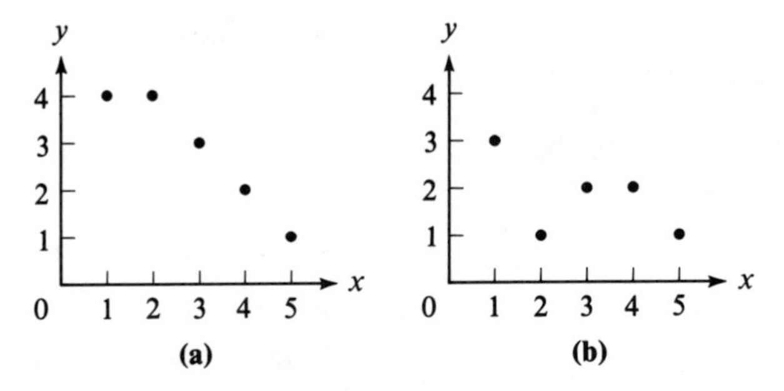

Figure A9.6.

Chapter 10

10.2 (a) In two throws, the probabilities for 0, 1, and 2 aces are 69.44, 27.78, and 2.78 percent, respectively. (b) In four throws, the probabilities for 0, 1, . . . , 4 aces are 48.23, 38.58, 11.57, 1.54, and 0.08 percent, respectively.

10.4
$$(p + q)^3 = \sum_{\nu=0}^{3} \binom{3}{\nu} p^\nu q^{n-\nu} = q^3 + 3pq^2 + 3p^2q + p^3.$$

10.6 Probability for any one patient to survive is $p = 0.2$, thus P(ν patients survive) $= b_{4,0.2}(\nu)$. (a) 41%. (b) 41%. (c) 18%.

10.7 40.2%, 40.2%, 16.1%, 3.2%, 0.32%, 0.01%.

10.9
$$\sigma_\nu^{\,2} = \overline{(\nu - \bar{\nu})^2} = \sum_\nu f(\nu)(\nu - \bar{\nu})^2 = \sum f(\nu)(\nu^2 - 2\bar{\nu}\nu + \bar{\nu}^2)$$
$$= \left[\sum f(\nu)\nu^2\right] - 2\bar{\nu}\sum f(\nu)\nu + \bar{\nu}^2\sum f(\nu) = \overline{\nu^2} - \bar{\nu}^2.$$

Note that if we replace the sums by integrals, we can prove the same result for a continuous distribution like the Gaussian.

10.10 For any p and q, $(p + q)^n = \sum\binom{n}{\nu}p^\nu q^{n-\nu}$. Differentiating twice with respect to p, we find

$$n(n - 1)(p + q)^{n-2} = \sum\nu(\nu - 1)\binom{n}{\nu}p^{\nu-2}q^{n-\nu}.$$

Multiplying by p^2 and setting $q = 1 - p$, we get

$$n(n - 1)p^2 = \sum(\nu^2 - \nu)b_{n,p}(\nu) = \overline{\nu^2} - \bar{\nu}.$$

Since $\bar{\nu} = np$, this implies that $\overline{\nu^2} = n(n - 1)p^2 + np$. Substituting into the result of Problem 10.9, $\sigma_\nu^{\,2} = \overline{\nu^2} - \bar{\nu}^2$ (with $\bar{\nu} = np$), we get the desired result.

10.13 9.68 percent (Gaussian approximation), 9.74 percent (exact).

10.14 $P(\nu \geqslant 18) \approx P_{\text{Gauss}}(\nu \geqslant 17.5) = P_{\text{Gauss}}(\nu \geqslant \bar{\nu} + 2\sigma) = 2.28$ percent.

10.16 $P(\nu \geqslant 12) = 0.65$ percent (if the fertilizer makes no difference). Thus 12 successes are "significant" and "highly significant."

10.18 Expect 360 passes; P(420 passes or more) $\approx P_{\text{Gauss}}(\nu \geqslant 360 + 5\sigma) =$ .00003 percent, which is highly significant.

Chapter 11

11.1 (a) For $\nu = 0, 1, \ldots, 6$, $p_{1/2}(\nu) = 60.7, 30.3, 7.6, 1.3, 0.2, 0.02, 0.001$ percent.

11.2 (a) $\sum p_\mu(\nu) = e^{-\mu}\sum\mu^\nu/\nu! = e^{-\mu}e^\mu = 1$.
(b) Differentiating (11.12) with respect to μ gives

$$\sum e^{-\mu}(\nu\mu^{\nu-1} - \mu^\nu)/\nu! = 0$$

or, using (11.12) again,

$$\sum \nu e^{-\mu}\mu^{\nu-1}/\nu! = 1.$$

Multiplying by μ we get $\sum\nu p_\mu(\nu) = \mu$, which is the desired result.

11.4 (a) μ = (number of nuclei) × p = 1.5. (b) The probabilities for ν decays, ν = 0, 1, 2, 3, are 22.3, 33.5, 25.1, and 12.6 percent, respectively. (c) $P(\nu \geqslant 4) = 6.5$ percent.

11.5 The vertical bars in Figure A11.5 show the observed distribution. The expected Poisson distribution, $p_3(\nu)$, has been connected by a continuous curve to guide the eye. The fit is good.

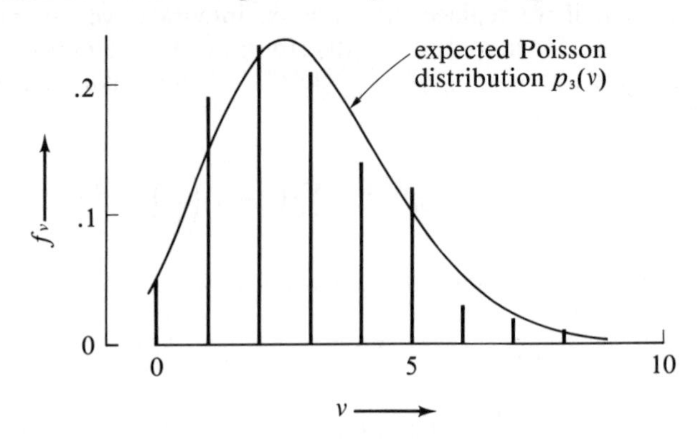

Figure A11.5.

11.7 $\bar{\nu} = 2.84$, $\sigma_\nu = 1.70$. These agree well with the expected values 3 and $\sqrt{3} = 1.73$.

11.9 (a) The probability of obtaining the observed count ν_o is $p_\mu(\nu_o) = e^{-\mu}\mu^{\nu_o}/\nu_o!$. The value of μ for which this is largest is found by differentiating with respect to μ and finding when the derivative is zero. The required derivative is

$$e^{-\mu}(\nu_o\mu^{\nu_o-1} - \mu^{\nu_o})/\nu_o!$$

and this is zero when $\mu = \nu_o$.

11.10 (a) 3.2 percent (Gaussian approximation), 3.4 percent (exact). (b) 8.5 percent (Gaussian approximation), 7.7 percent (exact).

11.11 (a) $p_9(7) = 11.7$ percent, etc. (b) $P(\nu \leqslant 6) + P(\nu \geqslant 12) = 40.3$ percent. A count that is as deviant as 12 is not at all surprising; so there is no reason to question that $\mu = 9$.

Chapter 12

12.1 Expected numbers = 7.9, 17.1, 17.1, 7.9, and $\chi^2 = 10$; the data fit the normal distribution very badly.

12.4 Expected numbers = 231.5, 138.9, 29.6; $\chi^2 = 2.5$; with three bins, $\chi^2 = 2.5$ is perfectly reasonable, and there is no cause to suspect the dice.

12.5 (a) In Problems 12.1 and 12.2, $c = 3$ and $d = 1$; in 12.3, $c = 1$ and $d = 5$; in 12.4, $c = 1$ and $d = 2$. (b) With ρ_{acc} known in advance, $c = 2$ and $d = 2$.

12.6 Since $d = 1$, $\tilde{\chi}^2 = \chi^2 = 10$; $P_1(\tilde{\chi}^2 \geqslant 10) = 0.2$ percent; so we can reject a normal distribution at 5 percent and 1 percent levels.

12.8 Probabilities for totals 2, 3, . . . , 12 are 1/36, 2/36, . . . , 6/36, . . . , 1/36. The expected numbers E_k are 10, 20, . . . , 60, . . . , 10. $\chi^2 = 19.8$, $d = 10$, and $\tilde{\chi}^2 = 1.98$. $P_{10}(\tilde{\chi}^2 \geqslant 1.98) = 3.2$ percent. At the 5 percent level we could say the dice are loaded, but not at the 1 percent level.

12.10 $\tilde{\chi}^2 = 1.2$. $P_2(\tilde{\chi}^2 \geqslant 1.2) \approx 30$ percent. Since $\tilde{\chi}^2 \geqslant 1.2$ is quite probable, there is no reason to suspect the dice.

12.12 (a) $E(v = 0) = 5.4$, $E(v = 1) = E(v = 2) = 10.8$, $E(v \geqslant 3) = 13.0$. $\chi^2 = 9.7$, $d = 3$, $\tilde{\chi}^2 = 3.2$. $P_3(\tilde{\chi}^2 \geqslant 3.2) \approx 2.5$ percent; so at the 5 percent level we would reject a Poisson distribution with $\mu = 2$. (b) $d = 2$, $\tilde{\chi}^2 = 0.3$, and the data are consistent with a Poisson distribution with $\mu = 1.35$.

12.13 (a) $\chi^2 = 2.25$, $P_1(\chi^2 \geqslant 2.25) \approx 14$ percent, and there is no significant difference. $P(v \geqslant 11) + P(v \leqslant 5) = 21.0$ percent.
(b) $\chi^2 = 6.25$, $P_1(\chi^2 \geqslant 6.25) \approx 1.2$ percent; $P(v \geqslant 224.5) + P(v \leqslant 175.5) = 1.4$ percent. At the 5 percent level there is a significant difference.
(c) Adjusted $\chi^2 = 1.56$. $P_1(\chi^2 \geqslant 1.56) \approx 21.2$ percent, in excellent agreement with the exact answer 21.0 percent.

Index

For definitions of important symbols and summaries of principal formulas, see the insides of the front and back covers.

Principal Formulas in Part II

Weighted Averages (Chapter 7)

If $x_1, \ldots, x_N$ are measurements of the same quantity x, with known uncertainties $\sigma_1, \ldots, \sigma_N$, then the best estimate for x is

$$x_{\text{best}} = \frac{\sum w_i x_i}{\sum w_i}, \qquad \text{(p. 150)}$$

where $w_i = 1/\sigma_i^2$.

Least-Squares Fit to a Straight Line (Chapter 8)

If $(x_1, y_1), \ldots, (x_N, y_N)$ are measured pairs of data, then the best straight line $y = A + Bx$ to fit these N points has

$$A = [(\sum x_i^2)(\sum y_i) - (\sum x_i)(\sum x_i y_i)]/\Delta,$$

$$B = [N(\sum x_i y_i) - (\sum x_i)(\sum y_i)]/\Delta,$$

where

$$\Delta = N(\sum x_i^2) - (\sum x_i)^2. \qquad \text{(p. 157)}$$

Covariance and Correlation (Chapter 9)

The covariance σ_{xy} of N pairs $(x_1, y_1), \ldots, (x_N, y_N)$ is

$$\sigma_{xy} = \frac{1}{N} \sum (x_i - \bar{x})(y_i - \bar{y}). \qquad \text{(p. 176)}$$

The coefficient of linear correlation is

$$r = \frac{\sigma_{xy}}{\sigma_x \sigma_y} = \frac{\sum (x_i - \bar{x})(y_i - \bar{y})}{[\sum (x_i - \bar{x})^2 \sum (y_i - \bar{y})^2]^{1/2}}. \qquad \text{(p. 180)}$$

Values of r near 1 or -1 indicate strong linear correlation; values near 0 indicate little or no correlation. (For a table of probabilities for r, see Appendix C.)